Plasma Fibronectin

HEMATOLOGY

Series Editor

Kenneth M. Brinkhous, M.D.
Department of Pathology
University of North Carolina
School of Medicine
Chapel Hill, North Carolina

Other Volumes in Preparation

Plasma Fibronectin

Structure and Function

edited by

Jan McDonagh

Beth Israel Hospital
and Harvard Medical School
and the Charles A. Dana Research Institute
Boston, Massachusetts

CRC Press

Taylor & Francis Group
Boca Raton London New York

CRC Press is an imprint of the
Taylor & Francis Group, an **informa** business

Library of Congress Cataloging in Publication Data
Main entry under title:

Plasma fibronectin.

(Hematology (New York, N.Y.) ; v. 5)
Includes bibliographies and index.
1. Fibronectins. I. McDonagh, Jan, [date].
II. Series. [DNLM: 1. Fibronectins. 2. Fibronectins--physiology. W1 HE873 v.5 / QU 55 P715]
QP552.F53P55 1985   612'.115   85-20436
ISBN 0-8247-7384-5

Series Introduction

Hematology as a field has followed a pattern for most of this century of major scientific discoveries, improved understanding of disease, and rapid applications of new knowledge in the clinic. Advances continue at an accelerating pace, so that all but the most zealous have difficulty in keeping up with the literature in even a limited area of specialized interest. As the explosive development of knowledge continues apace, it is a continuing challenge to keep abreast of significant new developments as they impact on clinical and laboratory hematology. The Hematology series is designed to help in this respect, by providing up-to-date and expert presentations of important subject areas in the field. It is hoped that these works, both individually and collectively, will become important

volumes to update one's information and for reference for the
clinician, investigator, teacher, and student, and in this manner
contribute to the advancement of hematology.

This volume of the series deals with a multifunctional plasma
and tissue protein, fibronectin, which participates in many significant biological and pathophysiological actions. As with many
major advances in hematology and allied fields, the beginning was
relatively obscure, with the identification of a new plasma protein
that was insoluble in the cold. This major discovery by John Edsall
and coworkers was described in 1948 in a paper whose title focused on their main concern, "The Separation of Purified Fibrinogen from Fraction I in Human Plasma." Relatively little attention
was given to this new plasma protein without known function in
the ensuing years. Then, in the last decade, it was rediscovered under other conditions and given other names. Fortunately, possible
confusion from a diverse nomenclature was short-lived, as Vaheri's
suggestion of the word "fibronectin" (fibra = fiber; nectere = to
connect) has been universally adopted. Fibronectin is now recognized as a macromolecule with many specific functional domains
which bind to fibrillary and particulate structures, as collagen and
fibrin, and to cells, as activated platelets and fibroblasts. Its molecular structure is well worked out, largely from studies on the fibronectin present in plasma. Plasma fibronectin deficiency appears
in a number of clinical states as sepsis, trauma, and shock. This
deficiency can be corrected with infusions of cryoprecipitate,
since fibronectin coprecipitates with other high-molecular-weight
plasma proteins, the factor VIII complex and fibrinogen. Tissue
fibronectin appears to be a ubiquitous component of tissues, present in the extracellular matrix and on cell surfaces. Tissue fibronectin is especially abundant in developing or newly formed tissues, as in wound healing or recently formed atherosclerotic
plaques. On the other hand, it is sparse in scars and fibrotic plaques.
There is much interest in the possible role of fibronectin in cell
transformation and its contributions to the invasive properties of
tumor cells.

The editor of this volume, Dr. Jan McDonagh, and her co-
authors have all contributed significantly to the development of
knowledge of fibronectin and are authorities on different aspects
of the rapidly growing field of fibronectin structure and function.
With the relevance of the fibronectin story to many aspects of

hematology, this volume presents a superb introduction to the subject, indicating the present state of the art, with particular emphasis on plasma fibronectin. With the participation of fibronectin in several seemingly unrelated physiological events, the full implications of this unique protein in medicine and hematology are probably just beginning to be realized.

Kenneth M. Brinkhous
Series Editor

Preface

From the discovery in the 1940s of cold-insoluble globulin, a plasma protein with no known function, through identification of fibronectin as a large, extracellular matrix protein, followed by recognition of cold-insoluble globulin as the soluble form of fibronectin, interest in the structure and function of fibronectin has continued to expand. In 1984 alone, more than 200 papers, ranging from basic studies to clinical applications of fibronectin, were published. When information is increasing so rapidly, it is difficult for even the most dedicated investigator to keep up to date. It almost seemed as though fibronectin changed suddenly from being a protein without a function to displaying a bewildering array of functions, becoming a participant in most of the reactions involving plasma proteins. Now it is becoming clearer which of these functions are likely to be physiologically important.

This volume focuses on plasma fibronectin and has two main aims. Basically, we wish to present the most current information concerning structure—both primary sequence as well as structural information obtained from electron microscopy—and other physical measurements. The second aim is to describe the functionally important interactions of fibronectin with other molecules, such as glycosaminoglycans, fibrinogen, and fibrin, and to assess critically its role in various processes, including platelet aggregation, fibrin formation, phagocytosis, and tissue repair. After a brief historical review (Chapter 1), Chapter 2 presents primary sequence data and delineates the interesting structural domains provided by the sequence information. Chapter 3 defines the structural properties of fibronectin as determined by electron microscopy, and Chapter 4 describes the physical and chemical properties of the molecule. Taken together, the structural information in these chapters provides the basis for analysis of the various functional domains of fibronectin. Chapters 5 and 6 describe the biochemical interactions of fibronectin with heparin and other glycosaminoglycans and with fibrinogen and fibrin. In the last four chapters, the roles of fibronectin in platelet activation and aggregation, fibrin formation, phagocytosis, and wound healing are examined.

As a compendium of current structure-function information, this volume will be useful to investigators involved in fibronectin research. Clinicians interested in learning about physiological and pathological relevance will find this a succinct summary of current data. For novice investigators just beginning research, this is a "user-friendly" introduction to a complex but fascinating subject. On reading the various chapters, the reader will see that there is much yet to be learned about fibronectin. The final purpose of this book is to stimulate further investigation.

Jan McDonagh

Contributors

David L. Amrani   Department of Medicine, Hemostasis Research Laboratory, University of Wisconsin Medical School, Milwaukee Clinical Campus, Mount Sinai Medical Center, Milwaukee, Wisconsin

Richard A. F. Clark   Department of Medicine, National Jewish Hospital, Denver, Colorado

Robert B. Colvin   Department of Pathology, Massachusetts General Hospital, Boston, Massachusetts

Harold P. Erickson   Department of Anatomy, Duke University Medical Center, Durham, North Carolina

Mark H. Ginsberg   Department of Immunology, Scripps Clinic
and Research Foundation, La Jolla, California

Masao Hada*   Department of Pathology, Beth Israel Hospital
and Harvard Medical School, and the Charles A. Dana Research Institute, Boston, Massachusetts

Jan Hermans   Department of Biochemistry, School of Medicine,
University of North Carolina, Chapel Hill, North Carolina

Helmut Hörmann   Department of Connective Tissue Research,
Max Planck-Institut für Biochemie, Martinsried bei München,
Federal Republic of Germany

Marek Kaminski   Department of Pathology, Beth Israel Hospital
and Harvard Medical School, and the Charles A. Dana Research Institute, Boston, Massachusetts

Gérard A. Marguerie†   Institut de Pathologie Cellulaire, Hôpital de Bicêtre, Bicêtre, France

Jan McDonagh   Department of Pathology, Beth Israel Hospital
and Harvard Medical School, and the Charles A. Dana Research Institute, Boston, Massachusetts

Michael W. Mosesson   Department of Medicine, Hemostasis Research Laboratory, University of Wisconsin Medical School,
Milwaukee Clinical Campus, Mount Sinai Medical Center, Milwaukee, Wisconsin

Torben E. Petersen   Department of Molecular Biology and Plant
Physiology, University of Aarhus, Aarhus, Denmark

Edward F. Plow   Department of Immunology, Scripps Clinic and
Research Foundation, La Jolla, California

Present affiliations:
*Department of Clinical Pathology, Tokyo Medical College, Tokyo, Japan
†INSERM, Centre d'Etude Nucleaire, Grenoble, France

Karna Skorstengaard   Department of Molecular Biology and Plant Physiology, University of Aarhus, Aarhus, Denmark

Livingston VanDeWater III   Department of Pathology, Beth Israel Hospital and Harvard Medical School, and the Charles A. Dana Research Institute, Boston, Massachusetts

Contents

1

Introduction

Jan McDonagh
Beth Israel Hospital
and Harvard Medical School
and the Charles A. Dana Research Institute
Boston, Massachusetts

Much of our basic knowledge concerning plasma proteins and their purification derives from work in the 1940s in the laboratory of Cohn at Harvard Medical School, where a group of talented investigators concentrated, for the first time, on the systematic fractionation of plasma [1]. Not surprisingly, plasma fibronectin was also identified by this group in a paper published by Edsall and colleagues in 1948 [2]. At first called cold-insoluble globulin because coprecipitation with fibrinogen was a major characteristic of the protein, plasma fibronectin did not appear to have any discernible function, nor did it attract much research interest. For some time there was little interest in the structure or function of this cold-insoluble globulin, and some investigators were not certain that it was a unique plasma protein. However, in 1968, Mosesson et al. discovered a patient with cryofibrinogenemia, in which the "cryofibrinogen" contained both fibrinogen and cold-insoluble globulin [3]. This observation led Mosesson and colleagues to purify the protein and to study its physical and chemical characteristics [4,5]. These studies clearly established cold-insoluble globulin as a unique plasma protein, present in significant concentration. However, it was as yet a plasma protein with no obvious function.

During the same period many research groups were characterizing an extracellular matrix protein secreted by normal fibroblasts in culture but not by transformed cells. This protein was given several names: large, external, transformation-sensitive protein (LETS protein) [6], surface fibroblast antigen [7], and cell surface protein [8], among others. Cold-insoluble globulin and LETS protein had similar molecular weights and appeared to have some similar physical characteristics. In 1975, Ruoslahti and Vaheri, using antiserum prepared to plasma cold-insoluble globulin, proved that soluble fibroblast surface antigen and plasma cold-insoluble globulin were immunologically identical [9]. An opsonic α_2-surface-binding glycoprotein in serum was later also found to be identical with cold-insoluble globulin [10]. These discoveries stimulated renewed interest in the plasma protein and led to creation of the term "fibronectin" [11]. Fibronectin is now the generally accepted term that describes a group of large glycoproteins with collagen binding activity, which are widely distributed both intracellularly and in extracellular matrix and in plasma. All molecular forms of fibronectin share antigenic expression and form precipitin lines of identity on double immunodiffusion against polyclonal antisera prepared to any fibronectin species.

Like the molecules themselves, research interests in fibronectin are now widespread and diverse, and it has become difficult to follow all areas at once. This monograph focuses on the structure and functions of plasma fibronectin with only modest attention to fibronectin that is cell associated or in the extracellular matrix. Although some differences depend on localization and physical state of fibronectin, plasma fibronectin serves as a useful model for all molecular forms. Current evidence indicates that there is one gene for fibronectin and that differences in the products secreted by different cell types are due to alternative splicing of fibronectin transcripts or posttranslational modifications [12-14]. Hence, the structure of plasma fibronectin is a prototype for all fibronectin molecules.

Circulating plasma fibronectin is synthesized and secreted by the liver [15]. The steady-state plasma concentration is about 300 μg/ml [4], or 0.7 μM. Platelets also contain fibronectin, which is synthesized in megakaryocytes. Over the last decade plasma fibronectin has been transformed from a protein in search of a function to a protein with possibly too many potential functions. The many possible functions of fibronectin derive from its specific binding domains, principally including fibrin, heparin, *Staphylococcus aureus*, collagen, and cell surface. The dimeric structure of fibronectin allows it to function as a molecular glue, holding various molecules together through its binding domains.

Plasma fibronectin contains two nearly identical polypeptide chains (M_r = 220,000-250,000), which are held together near the carboxyl terminus by two disulfide bonds. Structural and functional analyses, described in detail in this monograph, clearly indicate discrete functional domains in the molecule that correspond to distinct structural regions. Three types of primary sequence homology have been defined, and the various binding domains have been localized among the linear sequence. Current models of fibronectin describe it as a series of small, globular domains connected by short, flexible regions, somewhat analogous to beads on a string [16]. Under physiological conditions the molecule is maintained in a compact conformation through multiple specific interactions between the two polypeptide chains [17]. Binding to collagen converts fibronectin to a more extended form, which then enhances its ability to bind other molecules [18].

The amino-terminal region contains the binding domains for fibrin, proteoglycans, and *S. aureus*, as well as transglutaminase site. Next in the linear sequence is the collagen binding domain, followed by the cell attachment region, and finally additional proteoglycan and fibrin binding domains. With all these interactive regions, specific roles have been proposed for fibronectin in fibrin formation, platelet function, fibrinolysis, chemotaxis, phagocytosis, and opsinization. Current information relating to the functional roles of fibronectin is presented in this monograph. Suffice it to say by way of introduction that fibronectin is intimately involved in the processes of wound healing via its interactions with fibrin, collagen, and cells. Fibronectin is present in the fibrin clot and can be covalently cross-linked to it. Interactions between fibronectin and fibrin and between fibronectin and collagen may help to anchor the fibrin clot at the site of injury. Together, fibrin and fibronectin provide the scaffold for cellular migration into the wound site. Fibronectin is chemotactic for inflammatory cells, and it also has been implicated in the clearance of cellular debris by macrophages. Thus, fibronectin may have several important regulatory roles in tissue repair, ranging from formation and anchorage of the fibrin gel in the initial stage, through cellular infiltration, to removal of fibrin and cell debris in the later phases. Other specific functional roles for this highly interactive protein probably remain to be unraveled.

REFERENCES

1. J. T. Edsall, Some early history of cold-insoluble globulin. *Ann. N. Y. Acad. Sci. 312*:1-10 (1978).
2. P. R. Morrison, J. T. Edsall, and S. G. Miller, Preparation and properties of serum and plasma proteins. XVIII. The separation of purified fibrinogen from fractions of human plasma. *J. Am. Chem. Soc. 70*:3013-3108 (1948).
3. M. W. Mosesson, R. W. Colman, and S. Sherry, Chronic intravascular coagulation syndrome. Report of a case with special studies of an associated plasma cryoprecipitate ("cryofibrinogen"). *N. Engl. J. Med. 278*: 815-821 (1968).
4. M. W. Mosesson and R. A. Umfleet, The cold-insoluble globulin of human plasma. I. Purification, primary characterization, and relationship to fibrinogen and other cold-insoluble fraction components, *J. Biol. Chem. 245*:5728-5736 (1970).
5. M. W. Mosesson, A. B. Chen, and R. M. Huseby, The cold insoluble globulin of human plasma: Studies of its essential structural features. *Biochim. Biophys. Acta 386*:509-524 (1975).
6. R. O. Hynes, Alteration of cell surface proteins by viral transformation and by proteolysis. *Proc. Natl. Acad. Sci. USA 70*:170-174 (1973).
7. E. Ruoslahti, A. Vaheri, P. Kuusela, and E. Linder, Fibroblast suface antigen: A new serum protein. *Biochim. Biophys. Acta 322*:352-358 (1973).
8. K. Yamada and J. Weston, Isolation of a major cell surface glycoprotein from fibroblasts. *Proc. Natl. Acad. Sci. USA 71*:3492-3496 (1974).

9. E. Ruoslahti and A. Vaheri, Interaction of soluble fibroblast surface antigen with fibrinogen and fibrin. Identity with cold insolbule globulin of human plasma. *J. Exp. Med. 151*:497–501 (1975).
10. T. M. Saba, R. A. Blumenstock, P. Weber, and J. E. Kaplan, Physiologic role of cold-insoluble globulin in systemic host defense: Implications of its characterization as the opsonic α_2-surface-binding glycoprotein. *Ann. N.Y. Acad. Sci. 312*:43–55 (1978).
11. A. Vaheri and D. F. Mosher, High molecular weight, cell surface-associated glycoprotein (fibronectin) lost in malignant transformation. *Biochim. Biophys. Acta 516*:1–25 (1978).
12. H. Hirano, Y. Yamada, M. Sullivan, B. de Crombugghe, I. Pastan, and K. Yamada, Isolation of genomic DNA clones spanning the entire fibronectin gene, *Proc. Natl. Acad. Sci. USA 80*:46–50 (1983).
13. A. R. Kornblihtt, K. Vibe-Pedersen, and F. E. Baralle, Human fibronectin: Cell specific alternative mRNA splicing generates polypeptide chains differing in the number of internal repeats. *Nucleic Acids Res. 12*:5853–5968 (1984).
14. J. W. Tamkun, J. E. Schwarzbauer, and R. O. Hynes, A single rat fibronectin gene generates three different mRNAs by alternative splicing of a complex exon, *Proc. Natl. Acad. Sci. USA 81*:5140–5144 (1984).
15. M. R. Owens and C. D. Cimino, Synthesis of fibronectin by isolated perfused rat liver. *Blood 59*:1305–1309 (1982).
16. M. Rocco, M. Carson, R. Hantgan, J. McDonagh, and J. Hermans, Shape of the plasma fibronectin molecule in water and in glycerol-water mixtures, *J. Biol. Chem. 258*:14545–14549 (1983).
17. R. M. Robison and J. Hermans, Subunit interactions in human plasma fibronectin. *Biochem. Biophys. Res. Commun.* in press.
18. E. C. Williams, P. A. Janmey, J. D. Ferry, and D. F. Mosher, Conformational states of fibronectin; effects of pH, ionic strength, and collagen binding. *J. Biol. Chem. 257*:14973–14978 (1982).

2

Primary Structure

Torben E. Petersen and Karna Skorstengaard
University of Aarhus
Aarhus, Denmark

The suggested functions of fibronectin are all related to its binding
affinity to vastly different biological substances, especially cell surfaces. How is it possible for a single protein to bind negatively
charged compounds, such as heparin and DNA, and positively
charged compounds, such as polyamines, and how can it bind to
proteins with very different structures, such as collagen and fibrinogen? Within the last years a number of laboratories have obtained fragments from fibronectin containing one or more of the
specific binding activities, which suggests that fibronectin is a domain protein with the binding activities located in different domains. To provide a more precise characterization of the binding
domains of fibronectin, it is of obvious interest to determine the
primary structure of fibronectin preferentially on both the protein

level and the DNA level. In this chapter the available sequence
information of fibronectin is presented, based both on results obtained from human [1-5] and bovine [6-11] plasma fibronectin
and on results from recombinant DNA work on chicken genomic
DNA [12] and human [13, 14] and rat [14a] cDNA.

Fibronectin is a group of very similar proteins likely coded for
by only one gene [13, 14a]. Nevertheless, differences between fibronectin synthesized by cell cultures and fibronectin isolated
from plasma have been described. From a human cell line two different mRNA species were detected, and one of these coded for an
extra domain of 90 amino acid residues [14] when compared with
the other. Other suggested differences between the plasma and cell
fibronectin include carbohydrate [15], fragmentation patterns
[16], antibody specificity [17], and number of chains in the protein [18, 19].

Plasma fibronectin consists of two polypeptide chains linked
to each other by two disulfide bonds located near the carboxyl
terminus [9]. Each chain has a molecular mass of 220-250 kd but
shows a difference in mass of 10-25 kd when subjected to sodium
dodecyl sulfate polyacrylamide gel electrophoresis under reducing
conditions. The two chains have identical amino-terminal and carboxyl-terminal sequences, showing that the difference is not due
to limited proteolysis. By using proteolytic enzymes, an area near
the carboxyl terminus has apparently been found in one of the
chains only [20-22]. Different cDNA clones for rat fibronectin
have been isolated corresponding to three mRNAs [14a]. The differences between these mRNAs are found in positions that contain
a difference between the two chains of fibronectin [20-22]. Another difference between the two chains, probably of carbohydrate origin, has also been suggested near the gelatin binding domain [23].

FRAGMENTATION WITH PROTEOLYTIC ENZYMES

A huge number of fragments have been obtained from fibronectin
by using different proteolytic enzymes, but fewer of the fragments
have been characterized with respect to purification, amino acid
composition, and amino-terminal sequence. Nevertheless, these
studies show that most proteolytic enzymes are capable of hy-
drolyzing fibronectin in sensitive areas, and thrombin, for example,

an enzyme with rather restricted specificity, hydrolyzes fibronectin, at least in test tube experiments [2]. One aspect of fragments
of fibronectin is an enhanced activity in some assay systems when
compared with intact fibronectin. A 180 kdal fragment can augment ingestion of gelatin-coated particles by monocytes [24], a
transformation-enhancing activity has been associated with a 30
kd fragment that binds to gelatin [25], monocyte chemotaxis occurs with fibronectin fragments [26], and fragments of fibronectin promote DNA synthesis in fibroblasts [27]. In attempts to cor-
relate different disease states with the appearance or disappearance
of fibronectin (reviewed in Ref. 28), it is therefore not sufficient
to identify fibronectin by antibodies because these antibodies also
detect fragments of fibronectin. To obtain a meaningful correlation it is necessary to determine whether fibronectin appears as an
intact protein or as fragments. It is furthermore necessary to determine which sequence variant of fibronectin is present, as the different forms probably show variation in their binding characteristics. A number of fragments of fibronectin have been characterized from plasma cyroprecipitate [29] and in the plasma of certain cancer patients [30].

Fibronectin is covalently incorporated into the fibrin clot by
factor XIIIa [31] a reason for which it is of physiological interest
to characterize the fragmentation of fibronectin with plasmin.
Both human [32] and bovine [9] fibronectin are sensitive to plasmin, and the digestion pattern is similar but not identical [11].
The digestion of bovine fibronectin with human plasmin results in
four fragments with a molecular mass of 29, 170, 23, and 6 kd, in
the order they occur in the intact fibronectin molecule. The amino
acid sequence has been determined for the 29, 23, and 6 kd fragments and is described in detail in the next section. The 170 kd
fragment has been further digested with proteolytic enzymes, and
fragments binding to gelatin, DNA, cell surface, and heparin have
been isolated and the sequence determined or partly determined.

AMINO ACID SEQUENCES

The Amino-Terminal 29 kd Fragment

The sequence of the 29 kd fragment from bovine fibronectin has
been determined [9] and is shown in lines 1B–7B, Figure 1, resi-

Line No.   Type of Homology

```
 1  B  -    <EAQQIVQPQSPLTV → line 2
    H        EAQQMV

 2  B  I    SQCKPGSYDNGKHYQINQQWERTY-LGSAL-VCTCYGGSRG-FNCESKPEPE → line 3

 3  B  I    ETCFDK-YT-GNTYRVGDTYERPKDS--MIWDCTCIGAGRGRISCT-IA → line 4

 4  B  I    NRCHEG----GQSYKIGDTWRRPHETGGYMLECVCLGNGKGEWTCKPIA → line 5

 5  B  I    EKCFDQAA--GTSYVVGETWEKPY-QGWMMVDCTCLGEGSGRITCTSR → line 6

 6  B  I    NRCNDQDTR--TSYRIGDTWSKKDNRGN-LLQCICTGNGRGEWKCERH → line 7

 7  B  -    TSLQTTSAGSGSFTDVRTAIYQPQPHPQPPPY
    H        AAVYQPQPHPQPPPY → line 8

 8  B  I    GHCVTDS---GVVYSVGMQWLKTQ--GNKQMLCTCLGNG---VSCQE → line 9

 9  B  II   TAVTQTYGGNSNGEPCVLPFTYNGKTFYSCTTEGRQDGHLWCSTTSNYEQDQKYSFCTDH → line 10

10  B  II   TVLVQTRGGNSNGALCHFPFLYNNHNYTDCTSEGRRDNMKWCGTTQNYDADQKFGFCPMAAHE → line 11

11  B  I    EICTTNE---GVMYRIGDQWDKQHDMG-HMMRCTCVGNGRGEWTCVAYSQLR → line 12

12  B  I    DQCIVD----GITYNVNDTFHKRHEEG-HMLNCTCFGQGRGRWKCDPV → line 13

13  B  I    DQCQDSETR--TFYQIGDSWEKYLQ-G-VRYQCYCYGRGIGEWACQPLQTYPDT → line 14
    C        DQCQDSETR--TFYQIGDSWEKYVH-G-VRYQCYCYGRGIGEWHCQPL

14  B  III  SGPVQVIITETPSQPNSHPIQW

15  B                                                                 Try
    H                                                                 SD
                                                                      SD → line 16

16  B  III  KVPPPRDLQFVXVTDVKITIMWTPPESXVTGYXXDXIPVNLPGEXGQXLXVSXNXFA
    H        TVPSPRDLQFVEVTDVKVTIMWTPPESAVTGYRVDVIPVNLPGEHGERLPISXN

17  B  III                                   DLPSTATSVNIPDLLPGRKY
```

Figure 1   Amino acid sequence of parts of human (H), bovine (B), and chicken (C) fibronectin. Residues are indicated by the single-letter code: A, Ala; B, Asx; C, Cys; D, Asp; E, Glu; F, Phe; G, Gly; H, His; I, Ile; K, Lys; L, Leu; M, Met; N, Asn; P, Pro; Q, Gln; R, Arg; S, Ser; T, Thr; V, Val; W, Trp; Y, Tyr; Z, Glx; X, residues not determined. The amino terminus is a pyroglutamic acid; *, transglutaminase site; ED, extra domain; square, glucosamine-containing carbohydrate; hexagon, galactosamine-containing carbohydrate; Pla, plasmin; Chy, chymotrypsin; Try, trypsin; Pep, pepsin. All cystines and cysteines are underlined. The disulfide bridges are indicated for type I homology in line 2, for type II homology in line 9, and the two cysteines are in lines 18 and 28.

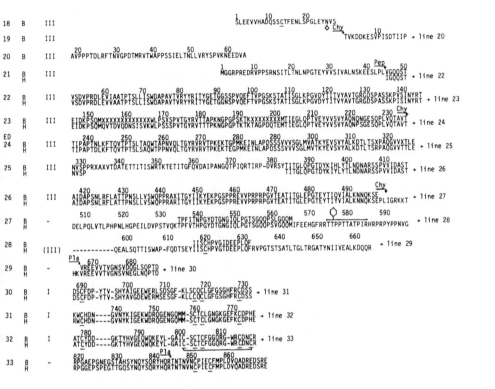

Figure 1 (continued)

dues 1-259. It has a pyroglutamic acid as amino terminus [6].
This was originally shown for intact human fibronectin by enzymatic liberation of the pyroglutamic acid [33], followed by the
sequence determination of the next five residues [2] (line 1H,
Fig. 1). The glutamine at position 3, line 1B, has been identified as
an acceptor site for factor XIIIa by incorporation of radioactive
putrescine into bovine fibronectin [6]. In human fibronectin an
acceptor site for factor XIIIa has also been found in the amino-terminal fragment [34, 35], and it is therefore likely that the glutamine at position 3, line 1H constitutes such a site. A second acceptor site for factor XIIIa has been localized near the carboxyl-
terminal part of human fibronectin [36], but the exact position is
not known. The first fourteen residues of the 29 kd fragment do
not show any homology to other parts of fibronectin, but the residues 15-242, lines 2B-6B, Figure 1, can be aligned as five domains
of mutually internal homology. This type of homology has been
named type I homology [9], and two other types of homologies
have been found in fibronectin [9], as discussed later. A type I
homology domain consists of about 45 residues and contains two
disulfide bridges. In each domain one of the bridges links half-
cystine 1 to half-cystine 3 and the other bridge is between half-
cystine 2 and half-cystine 4. In Figure 1 this bridge pattern has
been indicated in line 2B only.

The carboxyl terminus of the 29 kd fragment is arginine-259
shown in line 7B, Figure 1. This arginine is located in the middle
of a section of amino acid residues (line 7B, Fig. 1) sensitive to
proteolytic enzymes. The peptide bond between arginine-256 and
threonine-257 (alanine-257 in human fibronectin) can be hydrolyzed by trypsin [3], plasmin [9], and thrombin [2], and enzymes with specificities for other amino acids than arginine hy-
drolyze peptide bonds in this area also [1].

A number of binding interactions between the 29 kd fragment
and different biological compounds have been described. These include binding to fibrin [36], the surface of *Staphylococcus aureus* [37], actin [38], and heparin [39]. The binding to heparin is
sensitive to modulation by physiological concentrations of calcium ions [40].

The Gelatin-Binding 45 kd Fragment

The gelatin binding part of the 170 kd fragment can be obtained in a fragment of 45 kd after mild digestion with chymotrypsin [9]. The amino acid sequence of this fragment has been determined [10] and corresponds to residues 260–599 shown in lines 7B–14B, Figure 1. Residues 260–274 are part of a protease-sensitive segment, as mentioned. These residues have also been determined for human fibronectin [1, 3] and are shown in line 7H, Figure 1. The residues 275–313 can be aligned according to type I homology, and residues 314–436 correspond to two units of another type of homology called type II [9]. A domain of type II homology is about 60 residues long, with two disulfide bonds linking half-cystine 1 to half-cystine 3 and half-cystine 2 to half-cystine 4 [10]. In Figure 1 this bridge pattern in a domain of type II homology has been indicated in line 9 only. The two domains of type II homology are the only yet found in fibronectin and are in the sequence followed by three domains of type I homology corresponding to residues 437–577 (lines 11–13, Fig. 1). Finally, a segment of 22 residues (line 14) constitutes the carboxyl-terminal part of the 45 kd fragment with a tryptophan as the carboxyl terminus, in accordance with the amino acid specificity of chymotrypsin. These 22 residues have tentatively been aligned according to a third type of homology called type III, but more sequence information is necessary to settle this question.

A total of three carbohydrate prosthetic groups have been found in the 45 kd fragment linked to asparagine-399 in line 10, Figure 1, asparagine-497, and asparagine-511, both in line 12, Figure 1. All these groups are bound to an asparagine in the sequence asparagine-x-threonine, and glucosamine was found at all three positions when peptides containing the carbohydrate groups were hydrolyzed and subjected to amino acid analysis.

The only well-determined affinity attributed to the 45 kd fragment is the binding to gelatin. At present it is not clear which of the domains in the fragment are responsible for the affinity to gelatin. A 30 kd gelatin binding fragment has been isolated from human fibronectin [41]. As the amino-terminal sequences of the 45 kd fragment [9] and the 30 kd fragment [3] are identical, except for species-specific variations, it is possible that the 30 kd fragment does not contain the last two domains of type I homology, lines 12 and 13. When the sequence of the human 30 kd gelatin binding fragment becomes available, it will be interesting to compare the binding constant of the two fragments to gelatin. Unfortunately, we have not been able to isolate a 30 kd gelatin binding fragment from bovine fibronectin.

The Cell Binding 12 kd Fragment

After digesting a fragment of human plasma fibronectin with pepsin, a 12 kd fragment with affinity for the cell surface has been isolated [42], and the amino acid sequence determined [4]. This sequence corresponds to residues 45-152 in lines 21H-23H, Figure 1. A corresponding sequence (residues 1-151, lines 21B-23B) has also been obtained from bovine fibronectin [9] as a cyanogen bromide fragment 44 residues longer than the 12 kd human fragment. When the two sequences are compared, only four differences in positions 45, 61, 92, and 138 are found of 108 amino acid residues. This underlines the strong similarity between fibronectin from different species, which was earlier suggested by immunological cross-reaction experiments [43].

The sequence of the cyanogen bromide fragment shows internal homology, called type III homology [9]. A domain of type III homology is 90 residues long, and no disulfide bonds have yet been found in this type of homology. It is interesting that cell binding activity has been localized to a peptide corresponding to residues 123-152, lines 22H-23H, Figure 1 [44], a peptide only one-third of a type III domain.

The Heparin Binding 30 kd Fragment and Two Similar Fragments

When the 170 kd fragment was digested for 24 hr with a high concentration of chymotrypsin, three fragments were found that can be fractionated on a column of heparin-agarose [11], resulting in a pool with no affinity, a pool with weak affinity, and a pool with high affinity for heparin. From the pool with no affinity for heparin a fragment of about 40 kdal has been purified [11] and the amino-terminal sequence determined. This sequence corresponds to residues 1–64, lines 19–20, Figure 1, and belongs to type III homology. When this fragment was further degraded with cyanogen bromide a fragment beginning with glycine-2, line 21, was detected, indicating that the cell binding 12 kd fragment is included in the 40 kd fragment.

The pool with weak affinity for heparin contained fragments of about 30–40 kd and more than one amino terminus was detected. When the fragments were further digested with trypsin, a pure fragment of about 30 kd was obtained and the amino-terminal sequence corresponding to residues 1–59, lines 15B–16B, Figure 1, was determined. This sequence also belongs to type III homology, and a similar sequence has also been determined from human fibronectin, either as the amino-terminal sequence of a DNA-binding 24 kd fragment [3] or as the amino-terminal sequence of a weak heparin-binding 29 kd fragment [5]. This human sequence corresponds to residues 1–56, lines 15H–16H, Figure 1.

Finally, the pool with high affinity to heparin-agarose was eluted with about 0.3 M NH_4HCO_3 as a pure fragment. The amino acid sequence of this fragment has been determined [11] and corresponds to residues 235–497, lines 24B–26B, Figure 1. The sequence from bovine fibronectin was determined by isolating peptides from the bovine fragment, determining the sequence of these peptides, and then aligning the peptides on the basis of preliminary human cDNA sequence data [14, 45]. Part of the human sequence of the heparin binding fragment as predicted from the cDNA sequence [14, 45] is shown in Figure 1 and corresponds to lines 24H–26H. Other parts of the human sequence have been determined on the protein level [5] and correspond to residues 233–265 and 287–330, lines 23H, 24H, and 25H, Figure 1. When the sequence information of the heparin binding fragment determined on the protein level of human plasma fibronectin was compared with the sequence determined on the cDNA level a few differences appear, and the sequence shown in Figure 1 corresponds to the cDNA information. The heparin binding fragment is composed of three domains of type III homology, but how this fragment binds to heparin is not clear, and there is no apparent homology to the amino acid sequence of antithrombin III [46], another strong heparin-binding protein.

The Fibrin Binding 23 kd Fragment

This fragment consists of 178 amino acids, residues 666–843, lines 28–33, Figure 1. The sequence has been determined from bovine fibronectin on the protein level [9] and from human fibronectin on the cDNA level [13, 45], showing a total of 15 amino acid differences. The fragment contains three domains of type I homology, shown in lines 30–32, Figure 1, with a segment on the amino-terminal side (residues 666–686, line 29) and a segment on the carboxyl-terminal side (residues 818–843, line 33). Most of the sequence differences between the human and bovine 23 kd fragment are found in these connecting segments. The last domain of type I homology (line 32, Fig. 1) contains two "extra" half-cystines at positions 800 and 816, line 32, Figure 1, and because no sulfhydryl groups have been detected in the 23 kd fragment it is likely that these two half-cystines are linked to each other by a disulfide bond. Fragments similar to the 23 kd fragment from bovine fibronectin have been obtained from human [21, 22] and hamster [47] fibronectin, and the fragments were characterized by their affinity for fibrin. An interaction between the bovine fragment and fibrin-agarose has also been found [11].

The Carboxyl-Terminal 6 kd Fragment Containing Two Interchain Disulfide Bonds

This carboxyl-terminal fragment consists of two identical 26-residue peptides linked to each other by two disulfide bonds between half-cystines in positions 850 and 854, line 33, Figure 1, but it is at present not known whether the peptides are bridged in a parallel or antiparallel configuration. The sequence of the peptide corresponds to residues 844–869, line 33, Figure 1, and is identical for human and bovine fibronectin [9, 13]. The peptide ends on a glutamic acid, showing that this residue must be the carboxyl terminus of intact plasma fibronectin because of the amino acid specificity of plasmin. Additionally, the cDNA sequence has a stop codon after the glutamic acid, further showing that no proteolytic process occurs at the carboxyl terminus of fibronectin before it is secreted into the plasma. The bovine 6 kd fragment contains a phosphate group linked to serine-867, but the significance of this modification is not clear.

The Carboxyl-Terminal Part Deduced from cDNA

By recombinant DNA work, clones have been obtained containing cDNA corresponding to the carboxyl-terminal part of human, bovine, and rat fibronectin [13, 14a]. The cDNA sequence of the human clone has been determined [13, 45], and the deduced amino acid sequence, residues 383–869, is shown in lines 25H–33H, Figure 1. This sequence begins in the middle of the 30 kd heparin binding fragment and therefore covers the area (lines 27 and 28) between this fragment and the 23 kd fibrin binding fragment. The sequence following the heparin binding fragment (line 27) does not belong to any of the three types of homology; the sequence in line 28 can be aligned as a shorter version of a type III homology domain. Peptides from this part of bovine fibronectin have also been obtained [11] and are shown in lines 27B and 28B. Three different mRNAs have been characterized from rat liver [14a], showing that the amino acid sequence in line 27 might be shorter or longer than the version in Fig. 1. This is in agreement with expected positions of differences between the two chains of plasma fibronectin [20–22]. It is remarkable that the human sequence is different from all the three variants found in the rat, indicating that this part of fibronectin is highly variable, probably due to different splicing.

Cystines and Cysteines

A total of 60 half-cystines have at present been found in fibronectin: 20 in the amino-terminal fragment, 24 in the gelatin binding fragment, 14 in the fibrin binding 23 kd fragment, and 2 in the carboxyl-terminal 26-residue peptide. All half-cystines are located in domains of type I or type II homology, except for the inter chain bridges. The disulfide bridge pattern has been determined for the domains of type I homology in the 29 kd fragment [9], but the rest of the disulfide bonds in this type of domains have not been determined but are suggested on the basis of homology. The two disulfide bridges in a domain of type II homology have been determined for one of the domains (line 9, Fig. 1) only [10].

Fibronectin contains two sulfhydryl groups [48–50], which are accessible to labeling under denaturing conditions. After labeling the 170 kd fragment with radioactive iodoacetic acid in 6 M guanidine hydrochloride, a peptide containing carboxylmethylcy-

steine has been purified and the amino acid sequence determined
[8]. The sequence is shown in line 18, Figure 1, with a cysteine in
position 13, and clearly this peptide can be aligned according to
type III homology. A second carboxylmethylcysteine-containing
peptide has been purified after labeling intact fibronectin in 6 M
guanidine hydrochloride, followed by digestion with chymotrypsin [11]. The amino acid sequence of this peptide corresponds to
residues 616–631, line 28B, Figure 1, and is identical with a
stretch of sequence predicted from human cDNA except for a likely species difference (residue 624). This shows that the second cysteine is located in the area of fibronectin between the heparin
binding fragment and the fibrin binding 23 kd fragment.

The two cysteines in fibronectin are not found in the same
homologous position, but both occur in a stretch of sequence
showing part of type III homology. The two cysteines might be
important for the attachment of fibronectin to the cell surface matrix [48] and the multimer formation of the fibronectin chains
[51].

Carbohydrate and Phosphate

Fibronectin is a glycoprotein with about 5% carbohydrate [33].
At least five carbohydrate prosthetic groups have been found in
bovine fibronectin [8, 10-11]. Four groups contain glucosamine
and are linked to asparagine at positions 399, line 10, 497 and
511, line 12, and 25 line 18. The first three groups are in the gelatin binding 45 kd fragment [10], and the fourth is near one of the
two cysteines [8]. A potential attachment site for carbohydrate is
also located in this sequence, namely, asparagine-leucine-serine
18-20, line 18, but no carbohydrate was detected in the sequence.
A glycopeptide containing galactosamine has been isolated from
bovine fibronectin [11], corresponding to residues 574-584, line
27H. The predicted human sequence contains five threonines, but
the exact location of the carbohydrate group(s) in the peptide
from bovine fibronectin has not been determined. The peptide or-
iginates from the area of fibronectin where a difference between
the two chains has been suggested [14a, 20-22] and is perhaps
present in only one of the fibronectin chains.

The structure of three oligosaccharide groups from bovine
plasma fibronectin has been determined [52] and is of the bi-

antennary complex type. Further, the structure of oligosaccharide groups from both plasma [15] and cellular [53] hamster fibronectin has been determined and shown to be distinctly different from each other with respect to the linkage of sialic acid, the degree of sialylation, and the absence or presence of fucose. A galactosamine-containing glycopeptide with a higher molecular weight than the major unit has also been found in cellular hamster fibronectin [53].

In bovine plasma fibronectin a phosphoserine is located only three amino acids from the carboxyl terminus [9] corresponding to residue 867, line 33B. Phosphoserine has been found in human fibronectin [54], and fibronectin from transformed cells is more extensively phosphorylated than fibronectin from normal cells [55], but the exact position(s) has not been determined.

HOMOLOGY

Type I

A total of 12 domains of type I homology have at present been found in fibronectin, with 9 in the amino-terminal third part of fibronectin and 3 near the carboxyl terminus. More than a thousand amino acid residues separate these two areas of fibronectin. Each unit of type I homology contains two disulfide bridges, and the exact bridge pattern has been determined for all the bridges in the amino-terminal 29 kd fragment, except two. Because of the homology an identical bridge pattern is excepted in the other units of type I homology. In Figure 2, the 12 domains of type I homology have been aligned with each other, and gaps have been introduced by reason of homology. Amino acids are boxed if more than half of the residues in one position are identical. Using this criterion 19 of 43 positions are the most conserved. When the residues between the first and the last half-cystine in each domain are compared, the degree of identity varies from 20 to 55%, with an average of 39%. The highest amount of identity is found between domain 4 in the amino-terminal 29 kd fragment and domain 2 in the 23 kd fragment near the carboxyl terminus. The most conserved part of a type I homology is the half-cystines in the characteristic sequence -Cys-X-Cys-X-Gly-X-Gly-X-Gly-. Additionally,

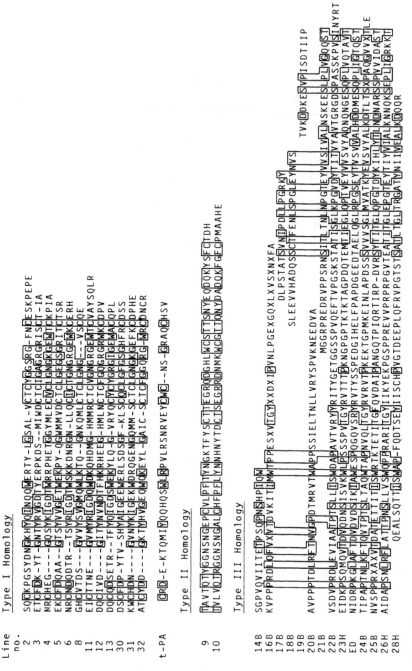

Figure 2

the two other half-cystines and a tyrosine are also found in very conserved positions.

In Figure 2 the amino-terminal 43 residues of the tissue plasminogen activator [56] are aligned [57], along with the 12 domains of type I homology, suggesting that this amino-terminal part of the activator constitutes a domain of type I homology. The homology of the activator sequence with fibronectin is not so marked, as only nine amino acids are correct when compared with the 19 most conserved residues in the 12 domains of type I homology. Unfortunately, the disulfide bridges have not been determined for the activator, but if the bridge pattern turns out as expected the homology is convincing. Activation of plasminogen by the tissue activator is accelerated by the presence of fibrin [58], an effect mediated through the binding of both the activator [59] and plasminogen [60] to fibrin. In contrast the activation of plasminogen by urokinase is not enhanced by fibrin, and when the amino acid sequences of the two activators [56, 61] are compared it is characteristic that urokinase lacks a domain of type I homology. The fibrin binding activity of the 29 and 23 kd fragments of fibronectin and the finding that the amino-terminal part of the tissue activator is responsible for at least some of the affinity to fibrin [57] indicate that these domains have evolved from a common ancestral fibrin binding structure.

Type II

Two domains of type II homology have been found in fibronectin, and both occur in the gelatin binding fragment, but it has not been proved that they are involved in the binding to gelatin. They show a strong mutual homology without any gaps, and the degree

---

Figure 2 Alignment of the three types of homology. Where the sequence is more extensively known from human fibronectin, this is shown; otherwise the bovine sequence is shown; H, human; B, bovine; ED, extra domain. The amino-terminal part of human tissue plasminogen activator t-PA [56] is aligned along with type I homology as suggested in Ref. 57. Residues are boxed if more than 50% of the amino acids in one position are identical. The single-letter code is shown in the legend to Figure 1.

of identity is 57%. The two disulfide bonds have been determined
for one of the domains of type II homology only. A domain of
type II homology has not yet been recognized in any other protein
than fibronectin.

Type III

From Figure 1 it is clear that at least 10 units of type III homology are present in fibronectin, and more are expected. The domains of type III homology show some characteristics, such as
lack of disulfide bonds, unusual amino acid composition, stability
against proteolysis, and a very well conserved size of 90 residues.
The lack of disulfide bonds is unusual, as most other proteins that
contain a number of domains also have disulfide bonds within the
domains. Two cysteines have been found in the domains of type
III homology, and they can both be labeled on the sulfhydryl
group under denaturing conditions. The content of histidine and
phenylalanine is low, and only one tryptophan is found in each
domain. Fragments composed of three to four units of type III
homology can be obtained when fibronectin is digested with a
variety of proteolytic enzymes, but it has been a problem to digest
these fragments further to determine their sequence. Even cyanogen bromide fragments show a high degree of stability against proteolysis. In Figure 2 the domains of type III homology are aligned
without introducing gaps, except for very few positions. The highest amount of homology is around the tryptophan in the amino-
terminal part and around a tyrosine in the carboxyl-terminal part,
but between-sequence variation is pronounced. The domain that
turns out to be most different from the general size is the last domain shown in line 28, which lacks residues both in the amino-terminal and the carboxyl-terminal ends. The domain containing the
cell binding sequence [44] is 94 residues long, and from the alignment in Fig. 2 approximately four "extra" amino acids are found
between residues 124 and 135 in line 22.

The difference between the two mRNAs found in a human cell
line [14] corresponds exactly to a domain of type III homology
called extra domain (ED), and this unit is located between line 23
and line 24 in Figure 1. The sequence is shown in Figure 2, but as

it has not yet been found in plasma fibronectin it is not included in
Figure 1 and it is possible that it is specific for cellular fibronectin.
From the protein sequence information of plasma fibronectin, it is
clear that at least one of the chains lacks ED because an overlapping
sequence without ED has been determined.

Genomic DNA

By recombinant DNA work and electron microscopy the entire
gene of chicken fibronectin has been visualized [12], showing that
it contains at least 48 small exons that are approximately 150 base
pairs long. This size corresponds nicely to polypeptide units of
type I and II homology. Furthermore, the DNA sequence was determined for exon 12, and the corresponding protein sequence is
shown in line 13C, clearly identifying exon 12 as the piece of
DNA coding for the last domain of type I homology in the 45 kd
gelatin binding fragment. The unit of type III homology is too big
to be coded for by a single exon, but it is possible that one domain
of type III homology is coded for by two exons. When the conserved tryptophan in the amino-terminal part is aligned with the
conserved tyrosine in the carboxyl-terminal part of a domain of
type III homology, some weak similarities can be seen. It is perhaps possible that, early in the evolution of fibronectin, a piece of
DNA corresponding to half a unit of type III homology (45 amino
acids) was duplicated and a type III homology domain formed.
After that event the DNA corresponding to the 90-residue piece
was multiplicated as found in the fibronectin molecule.

Although the size of the exons suggests that a single DNA segment of 150 base pairs is ancestral to the exons of the fibronectin
gene, it is clear from the sequence of the three types of homologies
that at least three pieces of DNA have been involved in the evolution of fibronectin. Perhaps the uniform size of the exons merely
reflects an evolution in units of nucleosome core particle. A nucleosome core consists of 140 base pairs of DNA bound to a his-
tone octamer in a stable configuration that can be crystallized
[62]. The DNA content of nucleosomes can vary from about 160
to 240 base pairs, and it is remarkable that many protein domains
fall in the range of 50–70 amino acid residues, corresponding to
the amount of DNA in a nucleosome.

MODEL OF FIBRONECTIN

The model of fibronectin shown in Figure 3 is based on more than
75% of the sequence, determined partly on the protein level partly
on the DNA level. Where results from both methods are available,
only very few discrepancies have appeared. The protein sequence
is determined on pooled material from many individuals, resulting
in an average structure, but the DNA technique results in data
from only one individual. A few variations have been found in the
sequence of bovine plasma fibronectin; for example, a methionine
was found in position 257, line 24, in about 10% yield, and a ser-
ine might be present in position 36, line 20B, in a small amount.

The amino acid sequence of fibronectin can be ordered in
three types of smaller units linked by pieces of unique sequence.
Some of these linking sequences are exposed in a way that makes
them sensitive to proteolysis and thereby facilitates the creation
of fibronectin fragments. Some well-defined fragments and their
binding affinities are indicated in Figure 3: 12 units of type I
homology, two units of type II homology, and 10 units of type III
homology are shown in Figure 3. Most of the remaining undeter-
mined sequence probably belongs to type III homology, resulting
in about 15 domains of type III.

An interesting aspect of the fibronectin structure is the possibility of sequence variations due to different splicing of the primary transcript. An extra domain of type III homology, called
ED, has been found in the mRNAs from a human cell line [14].
Evidence for this domain has not yet been seen in the plasma form
of fibronectin, and it is clear that it is absent in at least one of the
two chains of bovine plasma fibronectin. Another sequence variation has been found in the fibronectin mRNAs from rat liver
[14a], corresponding to a part of the protein that does not belong
to any of the three types of homology. This sequence contains a
relatively high number of prolines and is in Figure 3 symbolized
by a loop following the strong heparin binding domain. From this
area of bovine plasma fibronectin a galactosamine-containing peptide has been isolated. As predicted by the mRNA sequences from
rat liver, this peptide will not be present in all the synthesized
chains of fibronectin, suggesting that not only amino acid sequence variations are found between the two chains of plasma fibronectin but also variations in the content of galactosamine

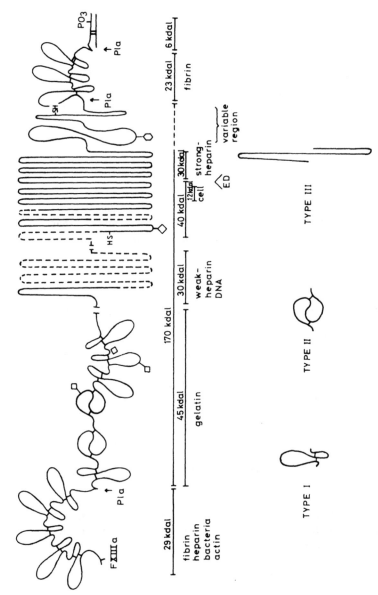

Figure 3 Model of a fibronectin chain. The 29, 170, 23, and 6 kd fragments were obtained by digestion with plasmin (Pla). The 45, 2 × 30, and 40 kd fragments were obtained by digestion of the 170 kd with chymotrypsin and the 12 kd fragment by digestion with pepsin. Square, glucosamine-containing carbohydrate; hexagon, galactosamine-containing carbohydrate; SH, cysteine; FXIIIa, transglutaminase site; PO₃, phosphate; ED, extra domain.

might be expected. Similarly, the possibility of other postsynthetic modification reactions, such as phosphorylation and crosslinking, might differ between the two chains. Finally, the human mRNA sequence is different from all the three variants found in rat liver, raising the possibility of species differences in this part of fibronectin in contrast to the rest of the molecule, where a very high degree of identity has been found between the amino acid sequences from different species.

REFERENCES

1. L. I. Gold, A. Garcia-Pardo, B. Frangione, E. C. Franklin, and E. Perlstein, Subtilisin and cyanogen bromide cleavage products of fibronectin that retain gelatin-binding activity. *Proc. Natl. Acad. Sci. USA* 76:4803-4807 (1979).
2. M. B. Furie and D. B. Rifkin, Proteolytically derived fragments of human plasma fibronectin and their localization within the intact molecule. *J. Biol. Chem.* 255:3134-3140 (1980).
3. H. Pande and J. E. Shively, NH_2-terminal sequences of DNA-, heparin-, and gelatin-binding tryptic fragments from human plasma fibronectin. *Arch. Biochem. Biophys.* 213:258-265 (1982).
4. M. D. Pierschbacher, E. Ruoslahti, J. Sundelin, P. Lind, and P. A. Peterson, The cell attachment domain of fibronectin. Determination of the primary structure. *J. Biol. Chem.* 257:9593-9597 (1982).
5. L. I. Gold, B. Frangione, and E. Pearlstein, Biochemical and immunological characterization of three binding sites on plasma fibronectin with different affinities for heparin. *Biochemistry* 22:4113-4118 (1983).
6. R. P. McDonagh, J. McDonagh, T. E. Petersen, H. C. Thøgersen, K. Skorstengaard, L. Sottrup-Jensen, S. Magnusson, A. Dell, and H. R. Morris, Amino acid sequence of the factor XIIIa acceptor site in bovine plasma fibronectin. *FEBS Lett.* 127:174-178 (1981).
7. K. Skorstengaard, H. C. Thøgersen, K. Vibe-Pedersen, T. E. Petersen, and S. Magnusson, Purification of twelve cyanogen bromide fragments from bovine plasma fibronectin and the amino acid sequence of eight of them. Overlap evidence aligning two plasmic fragments, internal homology in gelatin-binding region and phosphorylation site near C-terminus. *Eur. J. Biochem.* 128:605-623 (1982).
8. K. Vibe-Pedersen, P. Sahl, K. Skorstengaard, and T. E. Petersen, Amino acid sequence of a peptide from bovine plasma fibronectin containing a free sulfhydryl group (cysteine). *FEBS Lett.* 142:27-30 (1982).

9. T. E. Petersen, H. C. Thøgersen, K. Skorstengaard, K. Vibe-Pedersen, P. Sahl, L. Sottrup-Jensen, and S. Magnusson, Partial primary structure of bovine plasma fibronectin: Three types of internal homology. *Proc. Natl. Acad. Sci. USA* 80:137-141 (1983).

10. K. Skorstengaard, H. C. Thøgersen, and T. E. Petersen, Complete primary structure of the collagen-binding domain of bovine fibronectin. *Eur. J. Biochem. 140*:235-243 (1984).

11. K. Skorstengaard and T. E. Petersen, unpublished result, 1983.

12. H. Hirano, Y. Yamada, M. Sullivan, B. de Crombrugghe, I. Pastan, and K. M. Yamada, Isolation of genomic DNA clones spanning the entire fibronectin gene. *Proc. Natl. Acad. Sci. USA* 80:46-50 (1983).

13. A. R. Kornblihtt, K. Vibe-Pedersen, and F. E. Baralle, Isolation and characterization of cDNA clones for human and bovine fibronectins. *Proc. Natl. Acad. Sci. USA* 80:3218-3222 (1983).

14. A. R. Kornblihtt, K. Vibe-Pedersen, and F. E. Baralle, Human fibronectin: Molecular cloning evidence for two mRNA species differing by an internal coding for a structural domain. *EMBO J.* 3:221-226 (1984).

14a. J. E. Schwarzbauer, J. W. Tamkun, I. R. Lemischka, and R. D. Hynes, Three different fibronectin mRNAs arise by alternative splicing within the coding region. *Cell 35*:421-431 (1983).

15. M. Fukuda, S. B. Levery, and S. Hakomori, Carbohydrate structure of hamster plasma fibronectin. Evidence for chemical diversity between cellular and plasma fibronectins. *J. Biol. Chem. 257*:6856-6860 (1982).

16. M. Hayashi and K. M. Yamada, Differences in domain structures between plasma and cellular fibronectins. *J. Biol. Chem. 256*:11292-11300 (1981).

17. B. T. Atherton and R. O. Hynes, A difference between plasma and cellular fibronectins located with monoclonal antibodies. *Cell 25*:133-141 (1981).

18. K. M. Yamada, D. H. Schlesinger, D. W. Kennedy, and I. Pastan, Characterization of a major fibroblast cell surface glycoprotein. *Biochemistry 16*:5552-5559 (1977).

19. R. O. Hynes and A. Destree, Extensive disulfide bonding at the mammalian cell surface. *Proc. Natl. Acad. Sci. USA 74*:2855-2859 (1977).

20. H. Richter and H. Hörmann, Early and late cathepsin D-derived fragments of fibronectin containing the C-terminal interchain disulfide cross-link. *Hoppe-Seylers Z. Physiol. Chem. 363*:351-364 (1982).

21. K. Sekiguchi and S. Hakomori, Domain structure of human plasma fibronectin. Differences and similarities between human and hamster fibronectins. *J. Biol. Chem. 258*:3967-3973 (1983).

22. M. Hayashi and K. M. Yamada, Domain structure of the carboxyl-terminal half of human plasma fibronectin. *J. Biol. Chem. 258*:3332-3340 (1983).

23. T. Vartio, S. Barlati, G. DePetro, V. Miggiano, C. Stähli, B. Takács, and A. Vaheri, Evidence for preferential proteolytic cleavage of one of the two fibronectin subunits and for immunological localization distinguishing the cleavage. *Eur. J. Biochem.* 135:203-207 (1983).
24. J. K. Czop, J. L. Kadish, and K. F. Austen, Augmentation of human monocyte opsonin-independent phagocytosis by fragments of human plasma fibronectin. *Proc. Natl. Acad. Sci. USA* 78:3649-3653 (1981).
25. G. DePetro, S. Barlati, T. Vartio, and A. Vaheri, Transformation-enhancing activity of proteolytic fragments in fibronectin. *Proc. Natl. Acad. Sci. USA* 78:4965-4969 (1981).
26. A. Albini, H. Richter, and B. F. Pontz, Localization of the chemotactic domain in fibronectin. *FEBS Lett.* 156:222-226 (1983).
27. M. J. Humphries and S. R. Ayad, Stimulation of DNA synthesis by cathepsin D digests of fibronectin. *Nature* 305:811-813 (1983).
28. S. K. Akiyama and K. M. Yamada, in *Connective Tissue Diseases* (B. Wagner, R. Fleischmajer, and N. Kaufman, eds.). Williams and Wilkins, Baltimore, 1983, p. 55.
29. E. Ruoslahti, E. G. Hayman, E. Engvall, W. C. Cothran, and W. T. Butler, Alignment of biologically active domains in the fibronectin molecule. *J. Biol. Chem.* 256:7277-7281 (1981).
30. R. G. Parsons, H. D. Todd, and R. Kowall, Isolation and identification of a human serum fibronectin-like protein elevated during malignant disease. *Cancer Res.* 39:4341-4345 (1979).
31. D. F. Mosher, Cross-linking of cold-insoluble globulin by fibrin-stabilizing factor. *J. Biol. Chem.* 250:6614-6621 (1975).
32. F. Jilek and H. Hörmann, Cold-insoluble globulin. II. Plasminolysis of cold-insoluble globulin. *Hoppe-Seylers Z. Physiol. Chem.* 358:133-136 (1977).
33. M. W. Mosesson, A. B. Chen, and R. M. Huseby, The cold-insoluble globulin of human plasma: Studies of its essential features. *Biochim. Biophys. Acta* 386:509-524 (1975).
34. J. A. McDonald and D. G. Kelley, Degradation of fibronectin by human leukocyte elastase. *J. Biol. Chem.* 255:8848-8858 (1980).
35. D. F. Mosher, P. E. Schad, and J. M. Vann, Cross-linking of collagen and fibronectin by factor XIIIa. Localization of participating glutaminyl residues to a tryptic fragment of fibronectin. *J. Biol. Chem.* 255:1181-1188 (1980).
36. H. Richter, M. Seidl, and H. Hörmann, Location of heparin-binding sites of fibronectin. Detection of hitherto unrecognized transamidase sensitive site. *Hoppe-Seylers Z. Physiol. Chem.* 362:399-408 (1981).
37. H. Hörmann and M. Seidl, Affinity chromatography on immobilized fibrin monomer. III. The fibrin affinity center of fibronectin. *Hoppe-Seylers Z. Physiol. Chem.* 361:1449-1452 (1980).

38. D. F. Mosher and R. A. Proctor, Binding and factor XIIIa-mediated cross-
linking of a 27-kilodalton fragment of fibronectin to *Staphylococcus
aureus. Science 209*:927-929 (1980).

39. J. Keski-Oja and K. M. Yamada, Isolation of an actin-binding fragment of
fibronectin. *Biochem. J. 193*:615-620 (1981).

40. M. Hayashi and K. M. Yamada, Divalent cation modulation of fibronectin binding to heparin and to DNA. *J. Biol. Chem. 257*:5263-5267
(1982).

41. E. Ruoslahti, E. G. Hayman, P. Kuusela, J. E. Shively, and E. Engvall,
Isolation of a tryptic fragment containing the collagen-binding site of
plasma fibronectin. *J. Biol. Chem. 254*:6054-6059 (1979).

42. M. D. Pierschbacher, E. G. Hayman, and E. Ruoslahti, Location of the
cell-attachment site in fibronectin with monoclonal antibodies and proteolytic fragments of the molecule. *Cell 26*:259-267 (1981).

43. R. Ruoslahti and A. Vaheri, Interaction of soluble fibroblast surface antigen with fibrinogen and fibrin. Identity with cold insoluble globulin
of human plasma. *J. Exp. Med. 141*:497-501 (1975).

44. M. D. Pierschbacher, E. G. Hayman, and E. Ruoslahti, Synthetic peptide with cell attachment activity of fibronectin. *Proc. Natl. Acad. Sci.
USA 80*:1224-1227 (1983).

45. A. R. Kornblihtt, K. Vibe-Pedersen, and F. E. Baralle, personal communication, 1983.

46. T. E. Petersen, G. Dudek-Wojciechowska, L. Sottrup-Jensen, and S.
Magnusson, Partial primary sequence of antithrombin III, in *The Physiological Inhibitors of Coagulation and Fibrinolysis* (D. Collen, B. Wi-
man, and M. Verstraete, eds.). Elsevier/North Holland Biomedical Press,
Amsterdam, 1979, pp. 43-54.

47. K. Sekiguchi, M. Fukuda, and S. Hakomori, Domain structure of hamster plasma fibronectin. Isolation and characterization of four functionally distinct domains and their unequal distribution between two
subunit polypeptides. *J. Biol. Chem. 256*:6452-6462 (1981).

48. D. D. Wagner and R. O. Hynes, Domain structure of fibronectin and its
relations to function. *J. Biol. Chem. 254*:6746-6754 (1979).

49. M. Fukuda and S. Hakomori, Proteolytic and chemical fragmentation
of galactoprotein a, a major transformation-sensitive protein released
from hamster embryo fibroblasts. *J. Biol. Chem. 254*:5442-5450
(1979).

50. D. E. Smith, D. F. Mosher, R. B. Johnson, and L. T. Furcht, Immunological identification of two sulfhydryl containing fragments of human
plasma fibronectin. *J. Biol. Chem. 257*:5831-5838 (1982).

51. D. F. Mosher and R. B. Johnson, In vitro formation of disulfide-bonded
fibronectin multimers. *J. Biol. Chem. 258*:6595-6601 (1983).

52. S. Takasaki, K. Yamashita, K. Suzuki, S. Iwanaga, and A. Kobata, The
sugar chains of cold-insoluble globulin. *J. Biol. Chem. 254*:8548-8553
(1979).

53. M. Fukuda and S. Hakomori, Carbohydrate structure of galactoprotein a: A major transformation-sensitive glycoprotein released from hamster embryo fibroblasts *J. Biol. Chem. 254*:5451-5457 (1979).
54. M. H. Teng and D. B. Rifkin, Fibronectin from chicken embryo fibroblasts contains covalently bound phosphate. *J. Cell Biol. 80*:784-791 (1979).
55. I. U. Ali and T. Hunter, Structural comparison of fibronectins from normal and transformed cells. *J. Biol. Chem. 256*:7671-7677 (1981).
56. D. Pennica, W. E. Holmes, W. J. Kohr, R. N. Harkins, G. A. Vehar, C. A. Ward, W. F. Bennett, E. Yelverton, P. H. Seeburg, H. L. Heyneker, D. V. Goeddel, and D. Collen, Cloning and expression of human tissue-type plasminogen activator cDNA in *E. coli. Nature 301*:214-221 (1983).
57. L. Bányai, A. Váradi, and L. Patthy, Common evolutionary origin of the fibrin-binding structures of fibronectin and tissue-type plasminogen activator. *FEBS Lett. 163*:37-41 (1983).
58. P. Wallén, Activation of plasminogen with urokinase and tissue activator, in *Thrombosis and Urokinase* (R. Paoletti and S. Sherry, eds.). Academic Press, New York, 1977, pp. 91-102.
59. S. Thorsen, P. Glas-Greenwalt, and T. Astrup, *Thromb. Diath. Haemorrh. 28*:65 (1972).
60. S. Thorsen, Differences in the binding to fibrin of native plasminogen and plasminogen modified by proteolytic degradation. Influence of ω-amino carboxylic acids. *Biochim. Biophys. Acta 393*:55-65 (1975).
61. W. A. Günzler, G. J. Steffens, F. Otting, S-M. A. Kim, E. Frankus, and L. Flohe, *Hoppe-Seylers Z. Physiol. Chem. 363*:1155 (1982).
62. J. T. Finch, L. C. Lutter, D. Rhodes, R. S. Brown, B. Rushton, M. Levitt, and A. Klug, Structure of nucleosome core particles of chromatin. *Nature 269*:29-36 (1977).

3

Structure Seen by Electron Microscopy

Harold P. Erickson
Duke University Medical Center
Durham, North Carolina

Modern techniques of electron microscopy provide a simple and reliable visualization of protein structure at a resolution of 2-3 nm. In 1981, our group [1] and the laboratory of Engel et al. [2] independently published images of plasma fibronectin. The two studies were very largely in agreement in showing the molecule to be a long, thin strand with several or many points of flexibility. Additional details of the structure have been obtained by electron microscopy of large proteolytic fragments (domains) [1,3] and by microscopy of the molecule in different conformations [4]. Recently published data on the amino acid sequence [5] and gene splicing [6] suggest a more detailed structural model consistent with the shape and dimensions seen in the electron microscope images.

31

Most of the recent electron microscopic studies are in agreement on the basic structural model. There are some minor differences in dimensions, and different laboratories have emphasized
different details in the structure. Perhaps the most important feature still in question is the conformation of the molecule in solution and its precise relation to the shapes seen with different specimen preparation techniques. In this chapter I review the electron
microscopic studies, pointing out where there is agreement and at-
tempting to reconcile apparent contradictions or differences.

INTACT FIBRONECTIN

The most useful technique for visualizing fibronectin molecules is
the glycerol-drying, rotary shadowing technique, developed by
Shotton et al. [7] for the study of spectrin and by our group [8]
for the study of fibrinogen. The technique has subsequently been
applied to a wide variety of molecular structures, and the technical
aspects have been explored in some detail [9,10]. The reader is referred to these papers for discussion of the specimen preparation
method.

Figure 1 shows a field of fibronectin molecules at low magnification to demonstrate the range of structures found in these specimens. Selected molecules are shown at higher magnification in Figure 2. The molecules appear to be thin strands of uniform diameter, with no evidence for large globular domains. It is important
to note that the resolution of the images, about 3-5 nm, limits the
detail that can be seen. A globular domain of 50 kd would produce a nodule of 5 nm diameter. A nodular structure of this size
should be easily detectable if it occurred at a regular position on
the strand, but a domain much smaller than this would not be detected. The reader may note that nodular structures are seen on
some of the strands, but we have not been able to establish a pattern for their position except at the ends of the strands. These occasional nodular structures may be local kinks in the strand or ex-
traneous material stuck to the molecule. The very fine granularity
seen along the strands is due to the granular structure of the plati-
num shadow. The fibronectin molecule must, like all large proteins, have a well-defined tertiary structure, but the details of this
structure are just not revealed at the 3-5 nm resolution of these

Figure 1 A field from a specimen of rotary shadowed fibronectin demonstrating the variety of shapes of the molecule. The specimen was prepared
from a solution of protein in 0.2 M ammonium formate, 30% glycerol.
(×75,000) (Reprinted from Ref. 1 by permission of the publisher.)

images. A detailed model consistent with the dimensions seen in
these images is proposed in a later section.

The length of the molecules was determined by tracing the
molecule with a contour measuring device. There is some uncer-
tainty in these measurements, in that the contour is sometimes
ambiguous, and there is also some apparently real variation in
length. We measured 120 molecules and found a fairly uniform
distribution of lengths from 120 to 160 nm [1], with the average
140 nm. In a later study we measured an average of 135 nm for a
different set of images [4]. Engel et al. [2] obtained a somewhat

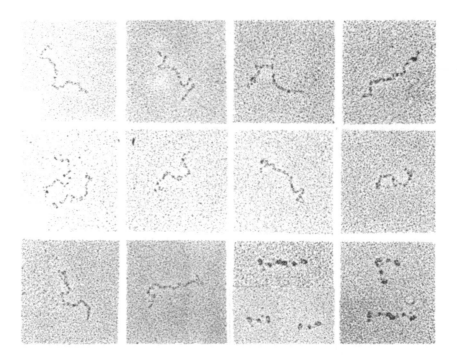

Figure 2 Selected examples of rotary shadowed fibronectin and, bottom row, two right-hand frames, fibrinogen molecules and end-to-end dimers that co-purified with the fibronectin. Characteristic structural features are the bend in the middle (first row, first frame) and the fork at the end of the strand (middle row, third frame). (× 150,000) (Reprinted from Ref. 1 by permission of the publisher.)

smaller value, 120 nm, for the average length. There are at least two possibilities for the difference. First, it may reflect a real difference in the length of the molecule, due to differences in the specimen preparation. Second, it may represent differences in the way the molecules are traced along the frequently ambiguous contour. Whatever the reason, the difference between lengths reported by these two groups is not large enough to be very significant.

   The thickness of the strand cannot be determined directly from these images. The apparent thickness of the shadowed mole-

cule depends very much on the way the metal coats the molecule
and can vary from one type of molecule to another [9]. A useful
estimate of the thickness can be arrived at indirectly. The density
of protein in compact domains is about 1.37 g/cc. If one assumes
that the fibronectin strand is a cylinder of uniform diameter,
packed with protein of this density, the diameter of this cylinder
can be calculated from the molecular weight (440,000), Avo-
godro's number, and the length. For a length of 140 nm this diameter is determined to be slightly over 2 nm. It is important to
note that this is actually the minimum possible diameter for the
strand, assuming that it is a uniform cylinder packed with protein.
If the strand has a bumpy structure, as in the string-of-beads model
presented in this chapter, the diameter of the bumps or nodules
will be larger.

   Although many of the molecules shown in Figure 2 are fairly
uniform strands, three structural features have been noted to occur with some consistency. One feature noted by Engel et al. is
a bend in the middle, which they estimate to average 70° both in
intact molecules and in proteolytic fragments comprising the middle section [2,3]. The first molecule, first row, of Figure 2 shows
this bend in its characteristic appearance. It should be noted, however, that many of the molecules in our images do not show this
bend. They are straight or slightly curved at the middle and sometimes show a strong bend off to one side of the middle. Another
feature we noted with some consistency at the middle of the
strand is the occasional segment of very low contrast, as if there
were a very thin or flat segment there. See, for example, the first
and third molecules of the top row, Figure 2. Finally, a prominent
feature that we have noted frequently at the ends of the strand is a
sharp bending to make a distinctive kink or fork. The third image,
bottom row, of Figure 2 shows this feature very strikingly, especially at the bottom where the "fork" is opened up. It is easy to imagine that the molecule in the next frame, which appears quite
short, simply has these forked ends tightly folded back into somewhat thicker domains. This region of the molecule is important because that is where the collagen binding domain is located.

   A more general structural feature, quite obvious from inspec-
tion of the images, is the flexibility of the strand. In independent
studies by our group and by Engel et al., the flexibility has been

interpreted somewhat differently. We concluded [1] that the strand was "extremely flexible" at many points along its length and did not believe we could identify hinge regions or points of pronounced bending. Engel et al. [2] suggested that the molecule had "limited flexibility" at a few points, corresponding to the hinge regions susceptible to proteolysis. They presented a histogram of bend frequencies along the length of the strand. The bend at 20 nm from the NH_2-terminal end and, to a lesser extent, the bend at 10 nm are quite convincing in this histogram. These points actually correspond to the forked structure that we identified at the end of the strand, so we agree that these are points of enhanced bending. For the rest of the molecule, from 25 nm to the middle of the strand, the histogram shows that the bend frequency is fairly uniform. This is consistent with our qualitative conclusion that bending can occur at many points along the length. In summary, the data from both laboratories support the interpretation that flexibility occurs, at many points along the strand, with one or two points of enhanced bending near the NH_2 terminus that can produce a forked structure at the end of the strand.

PROTEOLYTIC FRAGMENTS

Prior to visualization of the molecules by electron microscopy, some of the most suggestive evidence for structure was from the study of domains that could be isolated after limited proteolysis. These domains could be quite large, were relatively resistant to further proteolysis, and retained specific functional activities, such as collagen binding. An analogy to the structure of fibrinogen suggested that the domains might be large regions with a compact globular structure, connected by thin, flexible segments that would be susceptible to proteolysis. It was, in fact, our goal in beginning these studies to identify and map these globular domains. We were somewhat surprised to find that the molecule showed no prominent nodular structure.

The farily uniform diameter of the fibronectin molecule strongly suggests that the large domains isolated by proteolysis must be simply segments of the strand. Microscopy of the domains themselves has shown this to be true. A preparation of large collagen binding fragments, 180–210 kd as determined by sodium dodecyl sulfate (SDS) gel electrophoresis, had an average length of 60 nm, slightly less than half the length of the intact molecule [1]. The mass per unit length is essentially identical to that of the intact molecule. Several different proteolytic fragments have also been visualized by Odermatt et al. [3], with essentially the same results and conclusions: each fragment is a segment of the strand, of length proportional to its molecular weight. It should be emphasized that these fragments, although relatively resistant to further proteolysis, have the form of thin, flexible strands, and are not single domains with a compact globular structure.

The large proteolytic fragments retained 80–95% of the polypeptide chain, missing only the C-terminal segment where the two chains are linked by a disulfide to form the dimeric molecule. An important conclusion from the analysis of these fragments is that each polypeptide chain spans only one-half of the 140 nm long strand. Their C termini are joined in the center of the strand by a disulfide bond (and perhaps other, noncovalent interactions) and the NH_2 termini are at the ends. An alternative possibility, that the two polypeptide chains each spanned the entire 140 nm length of the strand in some kind of parallel or coiled structure, was eliminated by these observations.

Half-molecules of fibronectin, prepared by reducing the disulfide bond, have confirmed this structural interpretation. As visualized by electron microscopy (see Fig. 5) these indeed look like the fibronectin molecule cut in the middle. The average length of the half-molecules was determined to be 55 nm by Odermatt et al. [3] and 69 nm in our study [4], comfortably close to half the length of the intact molecules reported by the two groups. Preparations of half-molecules are potentially useful for a variety of functional studies, but it should be noted that the procedures for producing these half-molecules by reduction and alkylation are not completely straightforward. Examination of the preparation by electron microscopy is important to demonstrate that the strands remain unaggregated. The papers should be consulted carefully for the details of the procedures [3,4].

The structure of the plasma fibronectin molecule, drawn to scale as deduced from the electron micrographs of intact molecules, half-molecules, and proteolytic fragments, is shown in Figure 3.

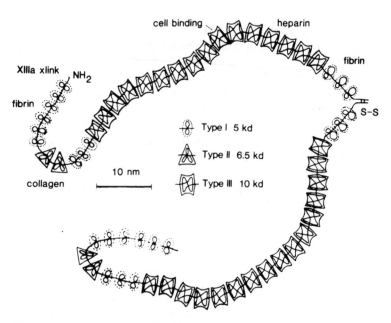

Figure 3 A scale model of the fibronectin molecule built from three types of small globular domains corresponding to the known sequence homologies. In overall length and diameter the model fits closely the structure visualized by electron microscopy. The detailed shape and folding depicted within each domain are for artistic emphasis only, since the higher resolution structure is not known.

FIBRONECTIN-FIBRINOGEN COMPLEXES

Our preparations of fibronectin frequently have a small but significant quantity of fibrinogen as a contaminating species. We actually preferred these mixed preparations for our electron microscope studies because the fibrinogen molecules are easily recognized and provide an internal structural marker. The contrast between the pronounced nodular structure of fibrinogen and the relatively thin, uniform strand of fibronectin, visualized side by side in the same field, was particularly striking (Figs. 2 and 4).

In addition to single fibrinogen molecules, we found many complexes that appeared to be fibrinogen dimers. These were identical in structure to end-to-end dimers that we had prepared in vitro by ligating the γ-chain site with factor XIIIa [11], and we

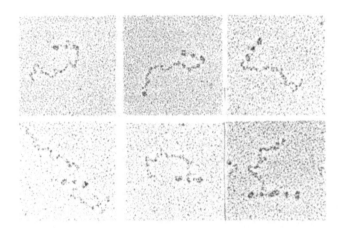

Figure 4 Complexes of fibronectin attached to fibrinogen molecules. The attachment site on fibronectin is always at the end of the strand. There are two
attachment sites on fibrinogen. In the top row the attachment is to the outer
nodule, where we have located the γ-chain ligation site. In the bottom row
the attachment appears to be at or near the central nodule. We have argued
that the C-terminal segment of the α chain, which contains the ligation site,
forms a separate nodule near the center of the molecule [12], which we propose is the attachment site in these complexes. (× 150,000) (From Ref. 1.)

concluded that these naturally occurring dimers were similarly
ligated. We also found frequent examples of fibronectin-fibrinogen complexes. As shown in Figure 4, these are of two types. In the
top row the fibronectin strand appears to be attached to the end
nodule of fibrinogen, at the point we have identified as the γ-chain
ligation site. In the bottom row of Figure 4 the fibronectin appears
to be attached to the middle nodule of the fibrinogen molecule.

    We have suggested [12] that the complexes of fibronectin at-
tached to the middle of fibrinogen may demonstrate ligation to
the α-chain site. This would require that the α-chain ligation site
of fibrinogen be near the center of the molecule. It has generally
been assumed that the C-terminal segment of the α chains, which
contain the ligation sites, are "free-swimming appendages" from
the end nodules. We have presented arguments that these chains
fold back toward the center of the molecule and form a fourth
nodule, which can be resolved in favorable images as a separate

structure close to the central nodule [12]. This is still a speculative proposal, but the observed attachment of fibronectin near the
center of fibrinogen adds substantial support.

There are several arguments that the observed complexes are
the result of covalent ligation by factor XIIIa. First, the specimens
are prepared at very dilute protein concentrations, on the order of
10^{-7} M. Noncovalent association of fibrinogen and fibronectin is
several orders of magnitude too weak to form complexes at the dilute protein concentrations we are studying. Thus, we can be con-
fident that the complexes are covalently ligated. Second, it is
known that factor XIIIa can cross-link fibronectin to fibrin [13],
and the cross-link acceptor site has been identified as a glutamine
residue three amino acids from the NH_2 terminus of the fibronectin chain [14]. It is remarkable that, in all the complexes, the attachment site on fibronectin is at the end of the strand, exactly
where we place the NH_2 terminus. Finally, in SDS gels of preparations rich in complexes, we have identified bands of about 300
kd, which would correspond to a covalent complex of fibronectin
with either an α or γ chain of fibrinogen (data not presented).

We have not yet isolated the ligated species and proven that
the ligation is to the γ-chain and α-chain sites. The structural identification of two distinct attachment sites is, however, a good argument that attachment can occur at both α- and γ-chain sites. Ligation of fibronectin to the α chains of polymerized fibrin has been
convincingly demonstrated [13]. In these experiments, however,
the γ-chain sites, which are faster reacting, were all consumed in
γ-γ dimers and were presumably unavailable for cross-linking to
fibronectin.

MODEL FOR THE STRUCTURE OF FIBRONECTIN: A STRING OF BEADS

The thin, flexible strand demonstrated by electron microsopy
seemed at first a bizarre structure for a protein molecule. In general, polypeptide chains tend to fold into compact globular domains, minimizing the contact of hydrophobic surfaces with
water. The electron micrographs show that there are no large globular domains (>5 nm) but domains smaller than 3 nm would not
be resolved by the shadowing technique. There are now several

lines of evidence suggesting that the fibronectin molecule does consist of a sequence of small, globular domains, similar to a string of beads.

Recent analysis of the amino acid sequence shows that fibronectin comprises repeating segments of homologous sequence [5, 6]. There are three types of homologies, designated Types I, II, and III, of sizes 5, 6.5, and 10 kd, respectively. A simple calculation shows that a 5 kd peptide could fold into a spherical domain of approximately 2.3 nm diameter, and a 10 kd peptide could form a 2.9 nm domain. These diameters are very close to the diameter of the fibronectin strand as seen by electron microscopy. Moreover, the total number of domains multiplied by their diameters gives a length reasonably close to the 80 nm observed for the most extended (half-molecule) strands. A "string-of-beads" model can therefore account for the structure seen in the electron microscope.

The model shown in Figure 3 is a synthesis of the amino acid sequence information arranged to fit the structure visualized by electron microscopy. There are several important features. (1) Each segment of Type I, II, or III sequence is assumed to fold into a separate globular domain. This assumption is compelling for Types I and II, which have internal disulfide bonds, and is reasonable for Type III. (2) The globular domains are separated by small links of polypeptide chains, giving potential points of flexibility. (3) The domains are shown as slightly oblate ellipsoids. This shape adjustment in the model gives a total length approximately equal to that seen in the electron microscope. The individual domains are too small to be resolved by conventional specimen techniques, in particular shadowing. Thus the slightly bumpy string of beads in the model is completely consistent with the relatively smooth strands visualized in our specimens.

EXTENDED AND COMPACT CONFORMATIONS

Alexander et al. [15] demonstrated some years ago that fibronectin underwent a change in conformation when the ionic strength was varied. The change was reflected by an increase in sedimentation coefficient from 10.8S in 0.4 M KCl to 13.6S in 0.04 M KCl. The conformation has been explored in more detail in recent studies using dynamic light scattering [16,17]. The nature of the

change in conformation could not be deduced until the structure
of the molecule was known, but once the molecule was visualized
in the electron microscope as a long, flexible strand a simple explanation was obvious: at high ionic strength the strand would be
more extended, and at low ionic strength it would be bent, curved,
or folded back into a more compact structure. In our original
study [1] we deliberately chose conditions of high ionic strength
that would favor the extended form; similar high ionic strength
was used by Engel et al. [2]. The molecules seen in these studies
were indeed well extended, the strand seldom crossing itself even
once (Figs. 1 and 2).

   We had noted more compact forms in some specimens in our
earlier work, and we subsequently began experiments to visualize
the compact form more consistently. Our initial attempts were not
very satisfactory in that specimens prepared from 0.02 M salt usually showed more extended molecules than compact forms. We
reasoned, however, that the salt must be concentrated considerably as the protein solution is dried in vacuum prior to shadowing.
Probably by the time the molecules were deposited on the mica
the ionic strength was sufficiently high to force the extended conformation. Consistent with this reasoning, we found that much
better specimens were obtained if the protein were dialyzed into
very low salt (1 mM buffer) and then diluted into glycerol solution
with no additional salt. In these specimens large fields of molecules in a compact form could be found reproducibly.

   Images of intact molecules and of half-molecules in the compact conformation are shown in Figure 5 and compared with

---

Figure 5   Electron micrographs of intact fibronectin (a,b), and half-molecules
prepared by reduction and alkylation (c,d) in high salt (0.2 M ammonium bi-
carbonate) and low salt (1 mM buffer). (a) Intact fibronectin, high salt. Characteristic V-shaped and straight molecules, as well as some proteolytic fragments of half the normal length, are seen in the extended conformation. (b)
Intact fibronectin, low salt. The strands are much more curved, bent, or
kinked to produce an apparently irregular tangle, the compact conformation.
(c) Half-molecules, high salt. The strands are half the length of the intact
molecules, and are relatively straight. (d) Half-molecules, low salt. Some
molecules are kinked and folded back on themselves; others are more uniformly curved into a circular form. (× 150,000) (Reprinted from Ref. 4 by
permission of the publisher.)

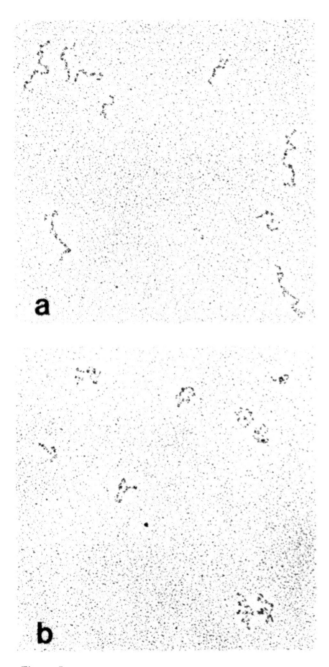

Figure 5

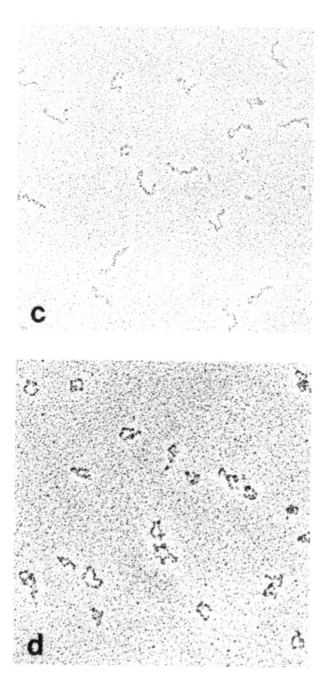

Figure 5 (continued)

images of the extended conformation. In specimens of intact fibronectin prepared at low ionic strength the strands are much more sharply curved and bent, crossing over themselves at one or more points to produce a tangled and more compact structure (Fig. 5b). There is no obvious regularity to the folding. Half-molecules also appear to be irregularly bent or kinked (Fig. 5d), and again there are no regular points of bending or of intrachain contact. In all cases it seems as if the strand tends to bend or kink at many points along its length in producing the compact conformation.

In an effort to quantitate the dimensions of the molecules at high and low ionic strength, we measured the "maximum extent," which was taken to be the distance across the molecule in the longest direction. Values of 100 and 50 nm were obtained for the intact molecules in the extended and compact conformations, respectively [4]. Price et al. [18] have presented images of intact fibronectin that appear identical to our compact conformation, although their specimen preparation conditions were rather different (discussed in the next section). Their compact molecules measured 50 X 30 nm in overall extent, the same as ours.

Having demonstrated the change in conformation by hydrodynamic measurements and having visualized the extended and compact forms by electron microscopy, one is faced with the important question of determining a mechanism for producing the two extreme conformations. Since the conformation is strongly affected by ionic strength it seems likely that electrostatic forces are involved. More specifically, one can suggest that the compact form is stabilized by electrostatic attractive forces between oppositely charged regions of the molecule, forces that will be enhanced at low ionic strength. The question, then, is how the oppositely charged domains are arranged on the strand. Two extreme possibilities have been suggested.

The first hypothesis is that distant segments of the strand, the NH_2-terminal ends and a region near the C termini, have a net positive and negative charge, respectively. It was proposed that, at low ionic strength, these distant segments would attract each other and cause the molecule to fold into a compact shape [16,19]. We proposed an alternate hypothesis [4] based on the model of the strand as a string of small globular domains. In this model we propose that adjacent small domains are oppositely charged and that electrostatic attraction at this very local level causes the strand to bend or kink at many points along its lengths.

We have determined [4], for the large segments whose amino acid sequence has been published [5], that most of the 5 kd domains have a substantial net charge, varying from 1 to 4 in magnitude. More importantly, the sign of the charge alternates from positive to negative as one passes from one domain to the next one in the strand. Thus, one would expect that neighboring domains would attract each other electrostatically. This attraction should be reduced in high salt, allowing the molecule to relax, and would be enhanced at low ionic strength. It is easy to imagine an arrangement of charges and shape of the domains that would produce a curvature or kinking of the strand as the domains are pulled together at low ionic strength.

An important conclusion from this model is that the strand would be more relaxed and flexible in the extended conformation. Fluorescent depolarization data of Williams et al. [16] have, in fact, demonstrated an increased mobility of the NH_2-terminal segment as the pH and ionic strength were increased. This observation is consistent with the prediction of our model, that the compact form is more rigid and the extended form more flexible.

The electron microscope images support our model in that the molecules appear to bend and kink at many points along the strand. The intact strands are generally tangled, with several points of contact or overlap. The points of contact do not appear to have regular positions, but it is difficult to trace the intact molecules to make this argument convincingly. The images of half-molecules are much simpler. The half-molecules show the same change in conformation, being much more curved and bent at low ionic strength than at high (Fig. 5c and d). The shorter strands are much easier to trace. It is clear that a large fraction have no crossovers, and for those that do the positions are not regular. This argues against an attraction of specific distant segments as the cause for the folding.

The most convincing evidence that the change in conformation occurs along short segments of the strand, rather than by attraction of distant segments, was provided by sedimentation of proteolytic fragments in high and low ionic strength [4]. Half-molecules showed a substantial increase in sedimentation from high to low ionic strength, similar to that of intact fibronectin. This demonstrates that the change in conformation occurs independently within each half of the molecule. More importantly, we found that large proteolytic fragments of 165-215 kd all had a pronounced ionic strength-dependent change in sedimentation coefficient, and a smaller fragment of 60 kd underwent a smaller but still significant change. These fragments all retained the collagen binding domain but were missing various portions of the C terminus of the chain, in particular the C-terminal segment supposed to be involved in the electrostatic attraction. That a fragment as small as 60 kd (about 10 small domains in the string-of-beads model) exhibits the change in conformation demonstrates that the conformation must be determined by short range interactions along the strand.

RELIABILITY OF THE ELECTRON MICROSCOPE

One is always concerned about artifacts in electron microscopy. Electron microscopy of fibrinogen was discredited for many years because some laboratories using the negative stain technique reported a globular structure, very different from the trinodular rod that had been seen by shadowing. The contradictory results were eventually resolved when the same trinodular structure was demonstrated by both techniques [8].

Koteliansky et al. [20] have presented images of shadowed fibronectin that they interpret as showing a 9 X 15 nm globular structure. There was no evidence for a strandlike substructure, and the globular particles they identified are actually much smaller than the compact form of fibronectin demonstrated by other groups. Their specimens were prepared at relatively high protein concentration, and their images have a very rough background. The identification of the globular particles as single fibronectin molecules is thus not convincing. Also, it is difficult to reconcile their images with the well-established strand structure for the molecule. Until this specimen preparation technique is tested more thoroughly it is difficult to accept these images as valid pictures of fibronectin.

We attempted to image fibronectin by negative staining in our earlier studies, but we had difficulty in demonstrating well-extended molecules. In our experience the prevalent form seen in

negative stain was similar to the compact conformation, even
when specimens were prepared at high ionic strength. The strandlike structure was usually obvious, but the strand was too coiled
and tangled to follow its contour. Extended molecules could be
found, but they were rare and not as clearly defined as the shadowed molecules. There is now reason to think that the compact
form visualized in these negatively stained preparations might correspond more closely to the 10S form obtained in moderate salt
conditions. In our earlier studies we preferred the shadowing technique because it gave well-extended molecules whose structure
could be easily interpreted. Engel et al. [2] did present a number
of convincing micrographs of extended molecules in negative stain.
The contour of the strand was usually not as clear as for the shadowed molecules, but the important point is made that the same
structure can be demonstrated by the two different specimen
preparation techniques.

    Price et al. [18] prepared shadowed specimens of fibronectin
that had been sprayed onto a carbon film, and they found almost
exclusively compact molecules. They were able to demonstrate extended molecules if they used polylysine as a substrate. They suggested that the substrate itself might affect the conformation of
the molecules as they are deposited. Unfortunately, they did not
state the salt concentration and other solution conditions used in
their specimen preparation, and they apparently did not try the
mica substrate that we have found much superior to carbon for
high-resolution shadowing. The suggestion that the substrate might
influence the conformation is still worth considering, even though
many of the differences might be due to the different extents of
salt concentration during drying. Mica and polylysine substrates
might induce the extended conformation, or the carbon film might
induce the compact form. The important question is how closely
the two shapes demonstrated by electron microscopy correspond
to the conformation in specified solution conditions.

    Rocco et al. [17] (see also Hermans, Chapter 4 of this volume)
have presented calculations of the Stokes radius for different conformations, based on a string-of-beads model with variable flexibility. Their Stokes radii can be converted directly to sedimentation
coefficients. For the molecule in an extended form, similar to the
shapes seen in Figures 2 and 5a, they calculate a value of 8S. For
the molecule in the most compact form that could be generated

by their model, similar in size and shape but perhaps less compact than those in Figure 5b, they calculated a value of 12S. Odermatt et al. [3] had previously used the Bloomfield method to calculate a sedimentation coefficient of 7.7S for a rigid molecule with the exact shape of an extended strand. Experimentally, a value of 10S is measured for the extended form in moderate solution conditions. A value of 8S is obtained only for extreme solution conditions: pH 11 [15,16] or very high salt and glycerol concentrations [17].

These calculations suggest that the extended form seen in the electron micrographs may correspond to the extremely extended 8S conformation, rather than to the 10S form found under more moderate solution conditions. Odermatt et al. [4] have already stated the same conclusion on the basis of the 8S value they obtained for their model of an extended molecule. This conclusion also makes sense in terms of the conditions for preparing the shadowed specimens. The molecules are sprayed onto the mica in a solution of 30% glycerol and 0.2 M salt, conditions that produce a moderately extended 10S conformation. As the solution is drying on the mica the water must evaporate first, leaving a solution near 100% glycerol and a salt concentration near 1 M. These conditions would indeed favor the extremely extended conformation, and it may be this form that is shown in our images. The moderately extended 10S conformation would then be somewhat more folded, but probably not as compact as the forms seen in Figure 5b. This interpretation could also explain our observation that negatively stained specimens showed mostly a compact form. In the negative stain technique the protein is not subject to drying and an increase in salt concentration, so the molecules might be visualized closer to their conformation in the starting solution.

The calculated sedimentation coefficient for the compact form, 12S, is less than the 13.5S measured at very low ionic strength. This suggests that the conformation in solution may be somewhat more compact than the 30 X 50 nm form seen in sha-dowed specimens (Fig. 5b) [14,18]. It should be noted, however, that the string-of-beads model used for the theoretical calculation [17] had a limited bending and could not easily achieve a form as compact as that seen in the micrographs.

In summary, electron microscopy has established convincingly that the fibronectin molecule is a long, thin strand with no large nodular domains. The C termini of two polypeptide chains are joined at the center of the strand, and the chains run separately to their NH_2 termini at the ends. A substructure in which the strand is composed of a string of small globular domains, which is suggested by the studies of the amino acid sequence and gene structure, is completely consistent with the structure and dimensions seen in the microscope. The strand has many points of flexibility and can change from an extended to a more compact conformation under different solution conditions. Although the exact relation of the images to the conformation in solution may still be questioned, micrographs of two extreme conformations have been obtained. These images demonstrate that the basic strandlike structure is retained, with the compact conformation produced by increased bending and kinking of the strand.

REFERENCES

1. H. Erickson, N. Carrell, and J. McDonagh, Fibronectin molecule visualized in electron microscopy: A long, thin, flexible strand. *J. Cell. Biol. 91*: 673-678 (1981).
2. J. Engel, E. Odermatt, A. Engel, J. Madri, H. Furthmayr, H. Rohde, and R. Timpl, Shapes, domain organizations and flexibility of laminin and fibronectin. *J. Mol. Biol. 150*:97-120 (1981).
3. E. Odermatt, J. Engel, H. Richter, and H. Hörmann, Shape, conformation and stability of fibronectin fragments determined by electron microscopy, circular dichroism and ultracentrifugation. *J. Mol. Biol. 159*:109-123 (1983).
4. H. P. Erickson, and N. A. Carrell, Fibronectin in extended and compact conformations—electron microscopy and sedimentation analysis. *J. Biol. Chem. 258*:14539-14544 (1983).
5. T. Peterson, H. Thøgersen, K. Skorstengaard, K. Vibe-Pedersen, P. Sahl, L. Sottrup-Jensen, and S. Magnusson, Partial primary structure of bovine plasma fibronectin: Three types of internal homology. *Proc. Natl. Acad. Sci. USA 80*:137-141 (1983).
6. J. Schwarzbauer, J. Tamkun, I. Lemischka, and R. Hynes, Three different fibronectin mRNAs arise by alternative splicing within the coding region. *Cell 35*:421-431 (1983).
7. D. Shotton, B. Burke, and D. Branton, The molecular structure of human erythrocyte spectrin. *J. Mol. Biol. 131*:303-329 (1979).

8. W. Fowler and H. Erickson, Trinodular structure of fibrinogen—confir-
   mation by both shadowing and negative stain electron microscopy. *J.
   Mol. Biol. 134*:241-249 (1979).
9. J. Tyler and D. Branton, Rotary shadowing of extended molecules dried
   from glycerol. *J. Ultrastruct. Res. 71*:95-102 (1980).
10. W. E. Fowler and U. Aebi, Preparation of single molecules and supra-
    molecular complexes for high-resolution metal shadowing. *J. Ulstruct.
    Res. 83*:319-334 (1983).
11. W. Fowler, H. Erickson, R. Hantgan, J. McDonagh, and J. Hermans,
    Cross-linked fibrinogen dimers demonstrate a feature of the molecular
    packing in fibrin fibers. *Science 211*:287-289 (1981).
12. H. P. Erickson and W. E. Fowler, Electron microscopy of fibrinogen, its
    plasmic fragments and small polymers. *Ann. N. Y. Acad. Sci. 408*:146-
    163 (1983).
13. D. Mosher, Crosslinking of cold-insoluble globulin by fibrin-stabilizing
    factor. *J. Biol. Chem. 250*:6614-6621 (1975).
14. R. P. McDonagh, J. McDonagh, T. Petersen, H. Thøgersen, K. Skorsten-
    gaard, L. Sottrup-Jensen, and S. Magnusson, Amino acid sequence of the
    Factor XIIIa acceptor site in bovine plasma fibronectin. *FEBS Lett. 127*:
    174-178 (1981).
15. S. Alexander, G. Colonna, and H. Edelhoch, The structure and stability
    of human plasma cold-insoluble globulin. *J. Biol. Chem. 254*:1501-1505
    (1979).
16. E. C. Williams, P. A. Janmey, J. D. Ferry, and D. F. Mosher, Conforma-
    tional states of fibronectin. *J. Biol. Chem. 257*:14873-14978 (1983).
17. M. Rocco, M. Carson, R. Hantgan, J. McDonagh, and J. Hermans, Shape
    of the plasma fibronectin molecule in water and in glycerol-water mix-
    tures. *J. Biol. Chem.* In press. (1983).
18. T. Price, M. Rudee, M. Pierschbacher, and E. Ruoslahti, Structure of
    fibronectin and its fragments in electron microscopy. *Eur. J. Biochem.
    129*:359-363 (1982).
19. H. Hörman, Fibronectin—mediator between cells and connective tissue.
    Klin. Wochenschr. 60:1265-1277 (1982).
20. V. E. Koteliansky, M. V. Bejanian, and V. N. Smirnov, Electron micro-
    scopy study of fibronectin structure. *FEBS Lett. 120*:283-286 (1980).

4

Physical Studies

Jan Hermans
School of Medicine
University of North Carolina
Chapel Hill, North Carolina

Various physical studies have been performed whose results tell us something about conformation and properties of the fibronectin molecule in solution. As is usual, these results at first were of a descriptive nature; however, as more information was collected, their interpretation in terms of a molecular model became possible. The results of these measurements have been particularly useful in that they clarified the interpretation of electron micrographs of fibronectin, and vice versa. Thus, one may now have considerable confidence in a description of the fibronectin molecule as a chain of independent globular domains connected by short flexible segments of polypeptide chain; the length of the molecule measured along this chain is about 140 nm, and its width varies around 2 nm. The overall shape of the molecule varies from an almost extended conformation to a compact form, depending on solvent conditions.

Many questions remain unanswered by these observations, which in fact raise several questions of their own. We do not yet understand the nature of the interactions between the domains and their macromolecular ligands. We understand incompletely the interactions between segments of a single fibronectin molecule that determine the molecule's shape, and we can only speculate about a possible physiological role of the observed change of the molecular shape with solvent conditions.

This chapter reviews results and interpretation of observations made using physical methods to study fibronectin in solution.

SIZE AND SHAPE OF THE MOLECULE

Molecular Weight

The sedimentation coefficient S of fibronectin at physiological ionic strength and pH has been determined to be around 13 $\times 10^{-13}$ sec^{-1} by measurements with the analytical ultracentrifuge.* The diffusion coefficient D has been determined with dynamic (or quasi-elastic) light scattering to be 2.25×10^{-7} cm^2 sec^{-1}. The molecular weight M of fibronectin can be obtained from the ratio of S and D with the equation

$$M = \frac{S}{D} \frac{NkT}{1 - \overline{v}\rho} \tag{1}$$

where

N = Avogadro's number
k = Boltzmann's constant (Nk = 8.31×10^7 erg K^{-1} mol^{-1})
T = absolute temperature
ρ = solvent density
$\overline{v}$ = the protein's partial specific volume

---

*Values of S and D have been corrected for changes in viscosity and density of solvent and are relative to water at 20°C.

The value of $\overline{v}$ of fibronectin has been determined to be 0.72 ml/g [3], the same value predicted from the amino acid composition and carbohydrate content [3,4]. M is found to be near 500,000 [1,2], that is, about 10% higher than twice the value of 220,000 per polypeptide chain estimated by polyacrylamide gel electrophoresis of reduced fibronectin in sodium dodecylsulfate solution. The origin of this relatively minor discrepancy is not yet known. Measurements of sedimentation equilibrium earlier gave a lower value of 450,000 [3]. A similarly elevated molecular weight was obtained from measurements of the intensity of scattered light [2].

Molecular Shape

Both the sedimentation and the diffusion coefficients of fibronectin may vary with solvent composition, without a change of molecular weight. These changes clearly correspond to a variation of the molecular shape. It is remarkable that considerable variation of S and D may occur *without* a significant change of spectra, which typically vary with changes in protein conformation and usually are sensitive indicators of transitions from a folded native conformation to a more disorganized denatured state (by circular dichroism and polarization of fluorescence from tryptophan side chains) [3].

Dependence of the sedimentation coefficient of fibronectin on pH and on ionic strength was first observed by Alexander et al. [4] (see also Ref. 5). Subsequently, Williams et al. [1] observed a similar dependence of the diffusion coefficient. These results and an additional observation at low ionic strength [2] have been plotted in Figure 1 as the corresponding (inverse) frictional coefficients f, computed with

$$D = \frac{kT}{f} \tag{2}$$

and Equation (1). Furthermore, Markovic et al. have recently reported a decreased value of S in the pH range 2 to 4 [37].

Figure 1 also shows the ionic strength dependence of diffusion coefficients measured in 30% glycerol [2]. The effect of urea on the diffusion coefficient is shown plotted in the same way in

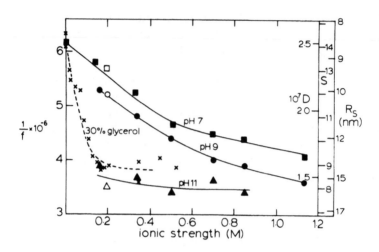

Figure 1   Frictional coefficients f of fibronectin as a function of ionic strength, at three different pH values and in 30% glycerol, pH 7.4. Open symbols mark values from measured sedimentation coefficients S [4]. Filled symbols mark values from diffusion coefficients D [1,2]. On the right are scales for D, S x 10^{13}, and Stokes radius R_S.

Figure 2 [see also Ref. 36]. These results were the first to indicate a transition to a *limiting* conformation with small S and D and large frictional coefficient. Obviously, the molecular conformation observed at pH 7 and low ionic strength is a more compact form than that observed under various less physiological conditions. It is not yet known if one is here dealing with an equilibrium between two well-defined conformations, or if one or many conformations intermediate between the observed extremes occur. The observed transition from compact to expanded form is reversed when urea or glycerol is removed by dialysis.

The compact form is presumably similar to native fibronectin as it circulates in plasma; in fact, it has recently been shown that purified fibronectin and native fibronectin in plasma behave identically in two experiments that measure molecular size: immunodiffusion and exclusion chromatography [6].

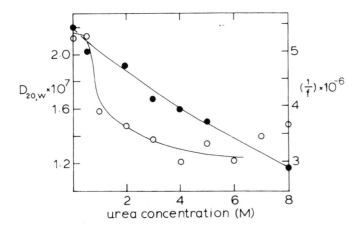

Figure 2 Dependence of the frictional coefficients f of fibronectin (from diffusion coefficients D) on urea concentration, at low ionic strength (●) and at ionic strength 0.35 M (○). (From Ref. 36.)

Quantitative Modeling

Traditionally, values of frictional coefficients and of the intrinsic viscosity $[\eta]$ have been used to estimate the size and shape of macromolecules in solution. Models of molecular size and shape that may be suggested on the basis of just one (or two) numbers are understandably poor of detail, and hence the success of molecular electron microscopy has long since doomed this approach to a secondary role. However, given the many artifacts that may arise in the process of sampling and specimen preparation for the electron microscope, it remains important to have an accurate and rapid technique (dynamic light scattering) that reports an average molecular dimension in unperturbed solution.

The relation between f and $[\eta]$, on the one hand, and molecular size, on the other, may be understood in terms of effective radii R, that is, the radii of spherical particles that would have the same f or $[\eta]$. These are found from equations due to, respectively, Stokes and Einstein:

$$f = 6\pi\eta_0 R_S \tag{3}$$

and

$$[\eta] = \frac{2.5v}{m} = 2.5 \times \frac{4\pi R_E{}^3}{3m}$$

where

η_0 = solvent viscosity
v   = molecular volume
m   = molecular mass   (4)

For a molecule with approximately spherical symmetry, R_S and R_E have similar values. This is in fact the case for fibronectin, both in the most compact and in the most extended form (Table 1).*

How do these values of R compare with the size and shape of the molecules observed in electron micrographs? This question

---

*More complicated equations for f and [η] of rodlike and ellipsoidal particles have been applied to model fibronectin [1], but this seems inappropriate given the coiled molecular shapes observed in electron micrographs.

Table 1 Effective Radii of Fibronectin from Diffusion Coefficient and Intrinsic Viscosity

Form	D ($\times 10^{-7}$ cm^2 sec^{-1})	[η] (ml/g)	R_S (nm)	R_E (nm)
Compact	2.5[a]		8.6	
	2.3[b]	10[b]	9.6	9.3
Extended	1.5[c]		14.5	
	1.4[d]	44[d]	15.5	16.0

[a]In 0.001 M phosphate buffer, pH 7.2 [2].
[b]At ionic strength 0.15 M, pH 7.4 [1].
[c]In 30% glycerol, 0.15 M NaCl, 0.1 M sodium phosphate, pH 7.2 [2].
[d]At ionic strength 1.0 M, pH 11 [1].

can be answered with some accuracy by use of a method developed by Bloomfield and coworkers [7,8], by which R_S (but
not R_E) of any model particle made up of spheres can be calculated. This method has been applied by Odermatt et al. [17] and
by Rocco et al. [2] in similar but not identical approaches. In both
cases, a chain of spheres (beads) was used to represent the fibronectin molecule, with the total volume that of the unhydrated, or
very slightly hydrated protein, and the length of the straightened
chain that suggested by the electron micrographs.

Odermatt et al. arranged the chains of beads according to the
shape of a number of molecules seen in their electron micrographs
[9] (see also Ref. 10), which had been obtained by sampling glycerol-water mixtures, calculated R_S for each, and obtained an average R_S very similar to that observed for the extended conformation.

Rocco et al. used a Monte Carlo method to generate models of
randomly flexed chains of beads and calculated an average R_S by
the Bloomfield approach. They thereby obtained a *family* of
models of different flexibility and accordingly different Stokes
radius R_S. Relatively extended models with calculated R_S near 15
nm, observed for the open form, had shapes very similar to those
of the molecules in electron micrographs obtained from glycerol-
water solutions of moderate ionic strength (Fig. 3a; see also Chap.
1, Fig. 5a) [9,10]; the average distance between the two ends of
these extended model chains was 80 nm, in good agreement with
the value of 90 nm estimated from the electron micrographs. The
most compact random chains had calculated Stokes radius near 10
nm, as observed for the compact form. The compact models, in
particular those in which the chain ends had been forced to interact closely with the middle part of th chain, are very similar in appearance to the more compact molecules observed more recently
in electron micrographs obtained with a variety of sampling proce-
dues (Fig. 3b and c; see also Chap. 1, Fig. 5b) [11-13].

The measured values of D and [η] constitute insufficient information on which to base selection of a model of the molecular
shape. In fact, the calculated Stokes radii of the string-of-beads
model are insensitive to details of the model. For example, introduction of a permanent kink of $70°$ in the middle of any model,
as suggested by electron micrographs [9], only barely lowers the

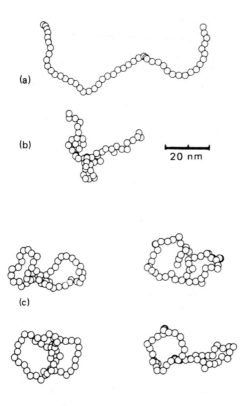

Figure 3 Examples of string-of-beads models of fibronectin. Model a is typical of stiff chains with the same average high frictional coefficient as observed
for the extended form of fibronectin. Model b is a highly flexed chain with
frictional coefficient slightly greater than observed for the compact form of
fibronectin. Models c are examples of chains with even lower frictional coefficients; the chains are highly flexed, and the two ends are bent back and
touch the chain in such a way that the two halves of the molecule cross one
another. (From Ref. 2.)

calculated Stokes radius. Similarly, a fully extended chain of beads
of length 140 nm has a Stokes radius of 16 nm, that is, barely
higher than the highest value observed experimentally. Even
though the difference is outside the accuracy of the diffusion coefficient determined with dynamic light scattering, it is the electron micrographs that provide conclusive evidence for the flexed
conformation of the extended form of fibronectin. Similarly, the

calculated Stokes radius of an extended double chain of beads of half the length (70 nm long) is 10.8, not much higher than is observed for the compact form. The latter arrangement of the beads seeks to represent a model of the compact form of fibronectin in which each half-molecule doubles back on itself [14] (see also Ref. 1). It is again the electron micrographs [11–13] that decide in favor of an irregular compact model, which may very well be stabilized by interactions between center and ends of the chain [2].

SPECTROSCOPY

Circular Dichroism

Circular dichroism (CD), the difference in absorption of left and right circularly polarized light, consists of bands that occur at the same wavelength as the bands in the absorption spectrum; the sign and the intensity of a band in the CD spectrum, relative to its intensity in the absorption spectrum, depend on the asymmetry of the environment of the chromophoric groups responsible for the absorption of the light. Strong CD bands for the peptide groups as chromophore are typically observed at wavelengths between 190 and 240 nm for proteins with a high content of α helix and/or β structure. Some proteins in addition show weaker bands at higher wavelengths for the aromatic side chains as chromophores. Circular dichroism is usually reported on a molar basis using the equation

$$[\theta] = \frac{M\theta}{10dc} \tag{5}$$

where

θ = ellipticity, the product of a constant factor and the observed difference in absorption
$[\theta]$ = molar ellipticity
d   = optical path (cm)
c   = concentration of protein (g/ml)

For M one may select either the molecular weight of the protein or the mean molecular weight of an amino acid residue, the latter the best choice in case the spectrum is due to peptide group absortion and reflects helical content, and the former preferable if a small number of chromophores is responsible for the CD spectrum.

The CD spectra of fibronectin have been measured in several laboratories and are in reasonably good agreement (Fig. 4) [3, 15-17]. Relative to that of most globular proteins, the spectrum is weak in the far ultraviolet (190-240 nm); this indicates that fibronectin's content of α helix and β structure must be very low.

The CD spectrum of fibronectin also contains bands in the range 260-300 nm; these are typical of the side chains of tyrosine and tryptophan inside a folded domain. The side-chain bands, although weak in an absolute sense, are relatively strong when compared with those of globular proteins in general. Since the bands between 260 and 300 nm are strong, one may expect additional relatively strong bands for the side-chain chromophores at lower wavelengths, that is, between 190 and 240 nm; the magnitude and sign of the additional bands are unpredictable. Thus the observed spectrum at lower wavelengths may actually be dominated by side-chain bands and contain an unknown (but smallish) contribution from helical polypeptide backbone.

Variation of the side-chain CD spectrum between pH 7 and 11 was exploited by Alexander et al. [4] to assign the band at 280 nm to tyrosine side chains. This band is not present at pH 11; tyrosine side chains become ionized when the pH is raised to 11, with a concomitant large change of the absorption spectrum and hence also the CD spectrum. On the other hand, the bands at 290 and 299 nm remain strong at pH 11; these may therefore be attributed to tryptophan. The CD spectrum of fibronectin has been found to vary with temperature at temperatures above $50°C$ due to unfolding of the folded protein domains (see the next section) [4].

Fluorescence

Measurements both of the intrinsic fluorescence of fibronectin, which is a property of the tryptophan side chains, and of the fluorescence of an attached dansyl probe have been made. In both cases measurements were made of polarization of the fluorescence, which reflects the ratio of the rotational relaxation time of the

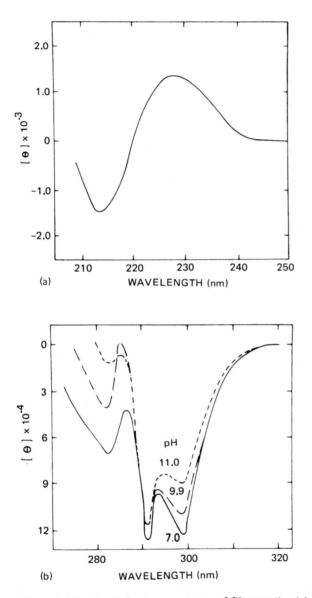

Figure 4 Circular dichroism spectrum of fibronectin. (a) Far-ultraviolet spectrum at pH 7 (as mean residue ellipticity). (b) Near-ultraviolet spectra at pH 7, 9.9, and 11 (as molar ellipticity). (From Ref. 4.)

fluorescent group to its fluorescent lifetime. The rotational relaxation time depends in an obvious way on the temperature and on
the viscosity of the solution; the rotational relaxation time of a
fluorescent group may be as large as that of the entire protein
molecule, or less if the group is loosely attached to the protein or
if the protein is flexible. The fluorescent lifetime is a parameter
that may be determined independently.

The fluorescent polarization of the tryptophan side chains was
found to be insensitive to an increase of the viscosity of the solvent. This means that the fluorescence lifetime is very short relative to the rotational relaxation time and that the latter cannot be
estimated from these measurements. Both the polarization and the
wavelength of maximum emission of the intrinsic fluorescence of
fibronectin were found to change at elevated temperature as the
protein unfolded (the next section) [4].

Dansyl groups have been covalently coupled at the amino terminal domain of fibronectin by reaction of the protein with dan-
sylcadaverine in the presence of the plasma transglutaminase factor XIIIa. From the temperature dependence of polarization between 8 and 38°C and the fluorescence lifetime, a rotational relaxation time of 77 nsec could be estimated [1]. This relaxation
time is much smaller than the value expected for an anhydrous
spherical protein with the same mass as fibronectin; hence it may
be concluded that the environment of the probe is flexible. This
may reflect a trivial flexibility of the spacer between probe and
protein. However, in this instance, the polarization was found to
decrease markedly when a ligand, collagen peptide CB7, was
bound to the protein [1]. The peptide binds to the second major
domain from the amino terminus of the chains, and the dansyl
probe is attached to the first domain [18]. Thus, one may conclude that the fluorescence polarization decreases when the ligand
is bound because the flexibility of the protein is greater with the
ligand bound.

Infrared Absorption Spectrum

The infrared spectrum of fibronectin has been reported [16]. The
spectrum in the region of the amide I band (1600-1700 cm-1) has
been corrected for contributions from side chains and decomposed

into four symmetrical bands. Comparison of the intensities and
positions of these bands with the known infrared spectra of α
helices and β structures suggests that the conformation of fibronectin consists for at most 35% of β structures and for an insig-
nificant amount of α helix.

CONFORMATIONAL TRANSITIONS

Change of Shape

Reversible variation of the conformation of fibronectin from a
compact to an extended form with relatively mild changes of
solution composition has been observed with electron microscopy
[11-13] and from measurements of the frictional coefficient as
described in the first part of this chapter. An extended conformation has been observed at high pH, at high ionic strength, in 30%
glycerol, in 1 M urea, and when certain collagen peptides with
known affinity for fibronectin (CB7 and CB6) are bound.

It is not unreasonable to assume that the same extended form
is observed under these different conditions. It has also been observed that the tryptophanyl CD bands at 290 and 299 nm are
relatively constant between pH 7 and 11 (Fig. 4) and that the
polarization and emission maximum of tryptophanyl fluorescence
and the CD bands at 299, 291, and 227 nm hardly change when
1.5 M guanidinium chloride, a stronger denaturant than urea, is
added to the solution [4]. From this one may conclude that extension of the molecular shape does not depend on change of the
conformation of folded domains but on changes of the interaction
between domains, which presumably are connected by intrinsically
flexible hinge regions. This shape change is reversible.

What might be possible physiological advantages for the existence of fibronectin in a compact circulating form that opens up
when a ligand is bound? It has been suggested [2] that perhaps
macromolecular ligands of fibronectin can be divided into at least
two categories on the basis of their relative affinities. A first class
of strong ligands, presumably including collagen, would be able to
bind to the circulating closed form of the protein, causing it to assume a more open conformation; however, these strong ligands are
presumably not encountered very frequently. The second class

would consist of weak ligands; these would not bind well to the
native form of fibronectin. Molecules frequently encountered by
fibronectin would be in this second category. If the weaker ligands
were, moreover, to bind well to the open form of the protein,
binding of a strong ligand and the consequent change of molecular
shape would result in enhancement of the affinity for the weaker
ligand. Two studies have reported evidence in favor of such an effect [19,20]. Although still quite speculative, this model is attractive for a protein whose supposed function is the linking of
dissimilar macromolecules.

What interactions cause the molecule to be compact under
some conditions and extended under others? A compact model of
fibronectin was first proposed by Hörmann [14]; in this model the
two polypeptide chains are folded back on themselves so that domains of opposite net charge are juxtaposed (Fig. 5a). This model
was also invoked by Williams et al. [1] to rationalize the values of
diffusion coefficient and intrinsic viscosity observed by them for
the compact form. Hörmann's model of fibronectin postulates in
addition a set of presumably more specific, complementary binding sites that can maintain intra- or intermolecular connections,
the former in the compact form and the latter in fibrils formed by
fibronectin in the presence of heparin [14] or polyamines [21]. It
may be noted that these same interactions might just as well cause
the two halves of the molecule to interact with one another as
cause each half to fold back itself (Fig. 5b) [2].

Erikson and Carrell [12] have observed that half-molecules
and fragments of fibronectin, just like the whole molecule, change
shape as a function of ionic strength. This could mean that the domains of opposite charge, whose mutual attraction gives the molecule a compact shape at low ionic strength, alternate much more
frequently than suggested by the model of Figure 5. (The models
of Figure 3 illustrate this behavior more correctly.) The (incomplete) amino acid sequence suggests that this is indeed the case
[12,22]. The observed sensitivity of the conformation of fibronectin to urea and to glycerol supports the idea that, in addition to
the nonspecific electrostatic interaction, a more specific interaction between one or two pairs of domains exists that also stabilizes
the compact conformation.

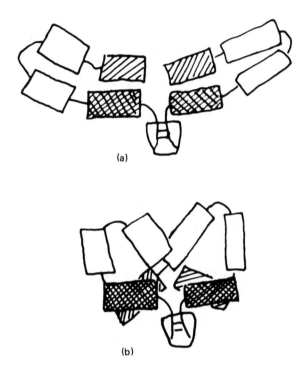

Figure 5 Models of the compact form of fibronectin. Differently shaded
blocks represent mutually attractive domains (which may have opposite
charge). In model (a), which was first suggested by Hörmann [14], the half-
molecule folds back on itself. Model (b) retains essential features of model
(a), but the arms of the molecule cross, and the molecule has a more irregular
shape, as suggested by recent electron micrographs. See also Figure 3.

Denaturation

Spectroscopic Studies

In contrast, major changes of these same spectroscopic characteristics have been observed at elevated temperatures and at higher
concentrations of guanidinium Cl [4,37]. At pH 7 a relatively sharp
transition occurs at 60°C, but at pH 11 a broader transition to an
apparently more progressively unfolded form is centered at 50°C.

At room temperature in guanidinium Cl solution unfolding begins at a guanidinium Cl concentration of 2 M; as the concentration is raised, the five selected spectroscopic parameters do not change proportionally. For some the change is complete when the concentration of the denaturant reaches 4 M, but others continue to change up to 6 M. This lack of parallelism is not unexpected for a protein consisting of a number of folded domains that are loosely connected.

Calorimetric Studies

Two groups have studied the thermal unfolding by measuring changes in heat capacity using scanning calorimeters. One group used a sensitive Privalov microcalorimeter [23], at a protein concentration of 1-1.5 mg/ml and scan rate of 1 K/min, and observed a peak with a maximum at 61°C and a total heat uptake of 1500–2500 kcal/mol, typical of protein unfolding [16]. The value calculated for total heat uptake depends on the position assumed for the baseline, which is very uncertain in this instance.

The other group used a modified commercial instrument, at protein concentrations of 10-45 mg/ml and a scanning rate of 10 K/min [24]. Since the sample was contained in a sealed container, measurements could be made to a temperature of 130°C. Three peaks of heat uptake were observed, a major one at 68°C, with heat uptake of 1200 kcal/mol, and two smaller ones at 85 and 120°C, with respective heat uptakes of 180 and 315 kcal/mol. (Observation of a higher temperature of 68°C, instead of 61°C, for the major transition is a result of the high scan rate.)

The second calorimetric study went somewhat beyond the other observations of unfolding, in that an ingenious attempt was made to identify the domains associated with the three transitions, in terms of loss of affinity for macromolecules. This was done via observation of irreversible loss, upon heating, of the activity of fibronectin as determined in various assays (agglutination of gelatin-coated latex beads, uptake of these same beads by liver slices, binding to heparin affinity columns, and immunological reactivity). The rates of these losses were compared with the rate constants for thermal denaturation of the domains, estimated from the dependence of the calorimetric results on the scan rate. The rate of loss of activities dependent on gelatin binding was found to be similar to the rate of unfolding of the least stable domains (transition at 60°C); when these are unfolded, immunological activity and heparin binding are unchanged. Part of the immunological activity is lost at a lower rate, similar to that at which the 83°C domain disappears; loss of the remainder of the immunological activity and of the heparin binding potential occurs considerably more slowly.

The question of the reversibility or irreversibility of the transitions observed at high temperatures and at high concentrations of denaturants has not been addressed or has been incompletely addressed in the reported calorimetric studies. Reversal of temperature-induced changes of the CD spectrum is very incomplete [37].

STUDIES WITH FRAGMENTS AND HALF-MOLECULES

The ease with which fibronectin can be broken down by digestion with numerous proteases to well-defined fragments is of itself strong evidence that the molecule consists of folded domains connected by flexible chain segments. The latter are, of course, much favored points of attack by proteases. One may separate the proteolytic fragments by chromatography over heparin and gelatin affinity columns, making use of the distribution of the affinities for different macromolecules over the domains.

By gentle reaction of fibronectin with mercaptans (for example, for 4 hr at 37°C with 20 mM dithioerythritol [17] one may reduce the interchain disulfide bridges and produce half-molecules of fibronectin that are thought to practically retain their native conformation. The newly produced sulfhydryl groups are alkylated with iodoacetate or iodoacetamide to stabilize the half-molecules.

Odermatt et al. [17] have studied five proteolytic fragments of different sizes and half-molecules by electron microscopy, sedimentation velocity, and CD spectra. The appearance of the fragments and half-molecules in the electron micrographs was found to be exactly what one would have expected from the electron micrographs of the intact molecule prepared in the same way. The observed frictional coefficients agreed well with those calculated for chains of beads configured according to the electron microscopic images. One should note that the electron microscope specimens were all sampled from glycerol-water mixtures, but the sedimentation rates were measured in 0.2 M NaCl, tris buffer at pH

7.4. As was mentioned, the frictional coefficient of fibronectin calculated by the same procedure does not agree with that observed in dilute buffered salt solution, being 50% higher, although equal to that observed in glycerol-water at moderate ionic strength. Thus the response to glycerol appears to be a property of the intact fibronectin molecule alone.

The CD spectra of fragments of fibronectin have been reported by two groups. The spectra all have a minimum near 210 and a maximum near 230 nm, as does that of the intact molecule (Fig. 4). In some fragments the band at 230 nm is more pronounced, and in others the band at 214 nm is the stronger. The sum of the spectra of the fragments equals the spectrum of the intact molecule [1]. More pronounced differences have been observed between the CD spectra of fragments at wavelengths near 280 nm; again, the spectra of the fragments approximately add to that of the intact molecule [15]. (See also Ref. 37.)

Fragments and half-molecules have been studied also by Erickson and Carrell [12], with electron microscopy and by sedimentation in a glycerol gradient. They found that the fragments and half-molecules assume compact forms at very low ionic strength, just as the whole molecule does, even in the presence of glycerol, but at higher ionic strength fragments and whole molecule are similarly extended (see the observed variation of the frictional coefficient with ionic strength in glycerol-water, Fig. 2).

Properties of derivatives obtained by reduction of fibronectin with mercaptans may differ due to difficulties of controlling the extent of the reaction and not permitting reduction of any of the numerous intrachain disulfides to occur. Thus, reaction with 10 mM dithiothreitol for several hours at 37° C was found to cause reduction of 80% of all disulfide bonds [25] and resulted in a decrease of affinity for gelatin or fibrin (although not heparin), and in enhanced susceptibility to intermolecular cross-linking by transglutaminase factor XIIIa. It is unfortunate that the extent of disulfide reduction of the half-molecules studied by the other authors is not known.

Data obtained by Robinson and Hermans [36] indicate that very mild reduction of fibronectin and subsequent alkylation of accessible sulfhydryl groups with iodoacetamide produces half-molecules that remain associated under conditions in which the whole molecule has a compact shape. These half-molecules dissociate and

can be separated from whole molecules by gel chromatography in
1 M urea (at moderate ionic strength). Light scattering measurements have shown that the mass and frictional coefficient of the
undissociated half-molecules is the same as that of native fibronectin. An average of 5 intrachain disulfide bonds per half-molecule was found to be in the reduced state (of a total of 30).

INTERACTIONS WITH MACROMOLECULES

Few physical studies of interactions of fibronectin with macromolecules have yet been reported. This is surprising, given the supposed biological function of fibronectin to bind to other biological
macromolecules and the very large literature describing observations of such interactions made with biochemical and biological
techniques.

The interactions of fibronectin and collagen have been studied
in greatest detail. The affinity of fibronectin for collagen is high.
One important biological function of plasma fibronectin may be
to facilitiate the adhesion of cells to collagen; indeed, attachment
of cells to a collagen substrate pretreated with fibronectin has
been used as an assay to quantitate binding of small amounts of
fibronectin to collagen [26].

A number of groups have reported absolute or relative affinities of fibronectin and fragments of fibronectin for different types
of collagen and its fragments. These studies required separation of
bound and unbound species, which was achieved by a variety of
techniques. (1) Either fibronectin or collagen was anchored to a
support, such as the bottom of a plastic dish or a polymeric bead
[19,27-30]. (2) Antibody was used to precipitate fibronectin together with other protein bound to it [31]. (3) The insolubility of
one of the components (collagen fibers or cellular fibronectin) allowed separation by centrifugation [32] or filtration [33]. These
methods have the advantage that they are rapid and can be designed to use very small amounts of protein. Their disadvantage is
that rinses required to remove all of that fraction of the protein
that is not bound to the substrate may remove some of the bound
protein as well. However, equilibration of collagen-fibronectin binding appears to be sufficiently slow that this problem is not serious.

The results of these studies can be summarized as follows. The
affinity of fibronectin is much higher for denatured collagen, with

a dissociation constant K_D near 10^{-8} M [26,31] than for native
collagen [30,31]. Incidentally, K_D for fibronectin-heparin interaction has been found to be of the same order of magnitude [33].
The fibronectin binding site of collagen is in a region sensitive to
digestion by collagenase [30] and has been localized to residues
757 to 791 of the $\alpha 1(I)$ chain [26,34]; the affinity of fibronectin
for a CNBr digest of the $\alpha 1(I)$ chain of collagen is only slightly less
than that for the undigested chain. Fragments of fibronectin obtained with elastase retain high affinity for denatured collagen;
fragments of molecular weight 60,000 and 40,000 were respectively
found to bind 5 and 25 times less strongly than intact fibronectin
[27]. On the other hand, the affinities of tryptic peptides having
molecular weights of 70,000 and 30,000 were lower by much
larger factors of 100 and 5000, respectively [28].

Two studies of interactions of fibronectin and collagen did not
employ physical separation of bound and unbound material. In
both studies changes of fluorescence anisotropy were used to mon-
itor binding, one using a dansyl label attached to fibronectin and
the other using a fluorescein label attached to denatured collagen
(gelatin).

The results of the first study [1] and the unexpected observation of an increase of probe mobility (decrease of fluorescence polarization) when collagen peptides CB7 or CB6 were bound have
been discussed. These measurements were not done to determine
dissociation constants of the complexes; however, one may esti-
mate $K_D \sim 5 \times 10^{-6}$ for peptide CB6 and $K_D < 3 \times 10^{-7}$ for the
more strongly binding peptide CB7 containing the high-affinity region identified by Kleinman et al. [34].

The study by Forastieri and Ingham [35] showed the expected
increase of fluorescence polarization of labeled gelatin due to loss
of probe mobility when fibronectin was bound. When unfraction-
ated commercial gelatin was used, the polarization was found to
change over a rather wide range of fibronectin concentration, as
expected if binding were possible at different sites varying greatly
in affinity. Separation of the gelatin according to molecular weight
produced fractions whose fluorescence polarization responded
more as expected for binding to a single class of sites, with the affinity greatest for the fraction of highest molecular weight. In fact,
the polarization of the fractionated labeled gelatin was found to
change *more rapidly* with the fibronectin concentration than ex-

pected for a simple binding equilibrium, as if there were some co-operativity, binding of one fibronectin molecule appearing to facilitate binding of another to the same gelatin molecule. One estimates a K_D of 10^{-9} for the interaction of high-molecular-weight gelatin and fibronectin. Interaction of the high-molecular-weight gelatin and a chymotryptic fragment of fibronectin with a molecular weight of 42,000 was found to give the same overall increase in the polarization of fluorescence. In this case the changes behaved according to a simple binding equilibrium; a Scatchard plot gave a K_D of 7×10^{-7} M.

REFERENCES

1. E. C. Williams, P. A. Janmey, J. D. Ferry, and D. F. Mosher, Conformational states of fibronectin: Effect of pH, ionic strength and collagen binding. *J. Biol. Chem.* 257:14973-14878 (1982).
2. M. Rocco, M. Carson, R. Hantgan, J. McDonagh, and J. Hermans, Dependence of the shape of the plasma fibronectin molecule on solvent composition: Ionic strength and glycerol content. *J. Biol. Chem.* 258:14545-14549 (1983).
3. M. W. Mosesson, A. B. Chen, and R. M. Huseby, The cold-insoluble globulin of human plasma: Studies of its essential structural features. *Biochim. Biophys. Acta 386:*509-524 (1975).
4. S. S. Alexander, G. Colonna, and H. Edelhoch, The structure and stability of human plasma cold-insoluble globulin. *J. Biol. Chem.* 253:1501-1505 (1979).
5. S. I. Miekka, K. C. Ingham, and D. Menaché, Rapid methods for isolation of human plasma fibronectin. *Thromb. Res.* 27:1-14 (1982).
6. M. Rocco and L. Zardi, Human plasma fibronectin shape in solution: Effects of NaCl concentration. Abstract, Meeting of American Society of Cell Biologists, San Antonio, Texas, 1983. *J. Cell. Biol.* 97:324a (1983).
7. V. Bloomfield, W. O. Dalton, and K. E. Van Holde, Frictional coefficients of multisubunit structures. I. Theory. *Biopolymers 5:*135-148 (1967).
8. J. Garcia de la Torre and V. Bloomfield, Hydrodynamic properties of macromolecular complexes. I. Translation. *Biopolymers 16:*1747-1753 (1977).
9. J. Engel, E. Odermatt, A. Engel, J. A. Madri, H. Furthmayr, H. Rohde, and R. Timpl, Shapes, domain organization and flexibility of laminin and fibronectin, two multifunctional proteins of the extracellular matrix. *J. Mol. Biol. 150:*97-120 (1981).

10. H. P. Erickson, N. Carrell, and J. McDonagh, The fibronectin molecule visualized by electron microscopy: A long, thin, flexible strand. *J. Cell. Biol.* *91*:673-678 (1981).

11. T. W. Price, M. L. Rudee, M. Piersbacher, and E. Ruoslahti, Structure of fibronectin and its fragments in electron microscopy. *Eur. J. Biochem.* *129*:358-363 (1982).

12. H. P. Erickson and N. Carrell, Extended and collapsed conformations of fibronectin visualized by electron microscopy. *J. Biol. Chem.* *258*: 14539-14544 (1983).

13. N. M. Tooney, M. W. Mosesson, D. L. Amrani, J. F. Hainfeld, and J. S. Wall, Solution and surface effects on plasma fibronectin structure. *J. Cell Biol.* *97*:1686-1692 (1983).

14. H. Hörmann, Fibronectin—mediator between cells and connective tissue. *Klin. Wochenschr.* *60*:1265-1277 (1981).

15. N. M. Tooney, D. L. Amrani, G. A. Homandberg, J. A. McDonald, and M. W. Mosesson, Near ultraviolet circular dichroism spectroscopy of plasma fibronectin and fibronectin fragments. *Biochem. Biophys. Res. Commun.* *108*:1085-1091 (1982).

16. V. E. Koteliansky, M. A. Glukhova, M. V. Bejanian, U. N. Smirnov, V. V. Filimonov, O. M. Zalite, and S. Y. Venyaminov, A study of the structure of fibronectin. *Eur. J. Biochem.* *119*:619-624 (1981).

17. J. Odermatt, J. Engel, H. Richter, and H. Hörmann, Shape, conformation and stability of fibronectin fragments determined by electron microscopy, circular dichroism and ultracentrifugation. *J. Mol. Biol.* *159*: 109-123 (1982).

18. R. P. McDonagh, J. McDonagh, T. Petersen, H. Thøgersen, K. Skorstensgaard, L. Sottrup-Jensen, and S. Magnusson, Amino acid sequence of the Factor XIIIa acceptor site in bovine plasma fibronectin. *FEBS Lett.* *127*: 174-178 (1981).

19. E. Pearlstein, Substrate activation of cell adhesion factor as a prerequisite for cell attachment. *Int. J. Cancer* *22*:32-35 (1978).

20. S. Johansson and M. Höök, Heparin enhances the rate of binding of fibronectin to collagen. *Biochem. J.* *187*:521-524 (1980).

21. M. Vuento, T. Vartio, M. Saraste, C. H. von Bonsdorff, and A. Vaheri, Spontaneous and polyamine-induced formation of filamentous polymers from soluble fibronectin. *Eur. J. Biochem.* *105*:33-42 (1980).

22. T. E. Petersen, H. C. Thøgersen, K. Skorstensgaard, K. Vibe-Pedersen, P. Sahl, L. Sottrup-Jensen, and S. Magnusson. Partial primary structure of bovine plasma fibronectin: Three types of internal homology. *J. Biol. Chem.* *254*:6054-6059 (1979).

23. P. L. Privalov, V. V. Plotnikov, and V. V. Filimonov, *J. Chem. Thermodyn.* *7*:41-47 (1975).

24. D. G. Wallace, J. W. Donovan, P. M. Schneider, A. M. Meunier, and J. L. Lundblad, Biological activity and conformational stability of the domains of plasma fibronectin. *Arch. Biochem. Biophys.* 212:515-524 (1981).

25. E. C. Williams, P. A. Janmey, R. B. Johnson, and D. F. Mosher, Fibronectin: Effect of disulfide bond reduction on its physical and functional properties. *J. Biol. Chem.* 258:5911-5914 (1983).

26. H. K. Kleinman, E. B. McGoodwin, and R. J. Klebe, Localization of the cell attachment region in types I and II collagens. *Biochem. Biophys. Res. Commun.* 72:426-432 (1976).

27. J. A. McDonald and D. G. Kelley, Degradation of fibronectin by human leukocyte elastase. *J. Biol. Chem.* 255:8848-8858 (1980).

28. E. Ruoslahti, E. G. Hayman, P. Kuusela, J. E. Shively, and E. Engvall, Isolation of a tryptic fragment containing the collagen-binding site of plasma fibronectin. *J. Biol. Chem.* 254:6054-6059 (1979).

29. D. F. Mosher, Fibronectin. *Prog. Hemost. Thromb.* 5:111-151 (1980).

30. E. Engvall and R. Ruoslahti, Binding of soluble form of fibroblast surface protein, fibronectin, to collagen. *Int. J. Cancer* 20:1-5 (1977).

31. F. Jilek and H. Hörmann, Cold-insoluble globulin (fibronectin). IV. Affinity to soluble collagen of various types. *Hoppe-Seyler's Z. Physiol. Chem.* 359:247-250 (1978).

32. H. K. Kleinman, C. M. Wilkes, and G. R. Martin, Interaction of fibronectin with collagen fibrils. *Biochemistry* 20:2325-2330 (1981).

33. K. Y. Yamada, D. W. Kennedy, K. Kimata, and R. M. Pratt, Characterization of fibronectin interactions with glycosaminoglycans and identification of active proteolytic fragments. *J. Biol. Chem.* 255:6055-6063 (1980).

34. H. K. Kleinman, E. B. McGoodwin, G. R. Martin, R. J. Klebe, P. Fietzek, and D. E. Woolley, Localization of the binding site for cell attachment in the α1(I) chain of collagen. *J. Biol. Chem.* 253:5642-5646 (1978).

35. H. Forastieri and K. C. Ingham, Fluid-phase interaction between human plasma fibronectin and gelatin determined by fluorescence polarization assay. *Arch. Biochem. Biophys.* 227:358-366 (1983).

36. R. M. Robinson and J. Hermans, Subunit interactions in human plasma fibronectin. *Biochem. Biophys. Res. Comm.* 124:718-725 (1984).

37. Z. Markovic, A. Lustig, J. Engel, H. Richter, and H. Hörmann, Shape and stability of fibronectin in solutions of different pH and ionic strength. *Hoppe-Seyler's Z. Physiol. Chem.* 364:1795-1804 (1983).

5

Interactions with Glycosaminoglycans

Michael W. Mosesson and David L. Amrani
University of Wisconsin Medical School
Milwaukee Clinical Campus
Mount Sinai Medical Center
Milwaukee, Wisconsin

This chapter is concerned with details of the interactions between fibronectin and a widely distributed class of macromolecules termed "glycosaminoglycans" that usually are covalently linked to a polypeptide core forming complex structures known as *proteoglycans*. The most salient characteristic of all forms of fibronectin is the presence on each of its polypeptide chains of discrete regions capable of binding to one or more of a variety of biological molecules (Fig. 1). These properties in turn confer upon fibronectin its most prominent biological function as an adhesive macromolecule mediating attachments between cells (such as fibroblasts, macrophages, and bacteria), collagen, fibrin(ogen), glycosaminoglycans (heparan sulfate), and/or other biological substances. In trying to understand the biological behavior of fibronectin con-

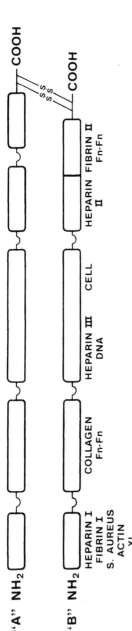

Figure 1 Schematic model of the plasma fibronectin molecule showing the location of certain of the well-characterized binding domains. The model represents a composite drawn from literature cited or summarized in the text and here. Abbreviations used are NH_2, amino terminus; COOH, carboxyl terminus; XL, factor XIIIa (plasma transglutaminase)-catalyzed cross-linking site; Fn-Fn, fibronectin self-assembly binding site; s-s, disulfide bridge; A, fibronectin A chain; B, fibronectin B chain. Other symbols are self-explanatory. The thin curved lines indicate protease-sensitive regions, cleavage at which yield the major proteolytic fragments containing the designated binding functions.

The NH_2-terminal domain has sites that bind to actin [1], *Staphylococcus aureus* [2], fibrin [3,4], and heparin, the latter by a calcium-sensitive mechanism [5]. Adjacent to this domain lies the carbohydrate-rich collagen-binding domain [6–11], which also contains a fibronectin self-association binding site [12]. Adjacent to this domain lies a cell binding domain [10,13] plus a weak heparin binding site [14] and a DNA binding site [15]. The remaining domains contain the heparin binding site of highest affinity (heparin II) as well as a second fibrin binding site (see text for details) plus a fibronectin self-association site [16] evidently complementary to the site in the collagen-binding domain. Finally, the dimeric chains of each molecule are held together by two disulfide bridges near the COOH terminus [11]. The A and B chains appear to differ from one another in that region of the molecule containing the heparin II and fibrin II binding sites (see text for details).

The heparin I (Ca^{2+} sensitive, intermediate affinity), and heparin II (high affinity, Ca^{2+} insensitive) sites have been characterized by several investigators (see text). A third binding site (heparin III) has weak affinity for heparin; its binding is inhibited by physiological concentrations of NaCl [14]. It is not certain whether this heparin binding site is the same as that described by Hayashi and Yamada [5].

cerned with binding to glycosaminoglycans, it is difficult and probably even undesirable to limit our considerations to just those specific interactions. Thus, although it is our intention to focus on structural and functional aspects of fibronectin-glycosaminoglycan interactions, we have tried to integrate these discussions, whenever warranted, to include consideration of other relevant binding functions of the molecule.

GLYCOSAMINOGLYCAN STRUCTURE AND PROPERTIES

Glycosaminoglycans (GAG) are complex polysaccharides that are found in animal tissues (Table 1). All GAG are polyanions with acidic sulfated or carboxylated groups, or both. Many of their functional properties are attributable to their polyelectrolyte nature. Except for keratan sulfate, their carbohydrate backbone consists of alternating uronic and hexosamine residues; all but hyaluronate are modified during biosynthesis to yield sulfated derivatives.

There is general agreement that most GAG (except for hyaluronate) in their native state are covalently linked at the reducing terminal sugar to a polypeptide core [17,18]. These protein-polysaccharide complexes are proteoglycans (PG).

Owing to their structural properties, GAG interact with many different biological macromolecules [17], including fibronectin. In this regard, it is important to note that most GAG also bind to collagen at physiologic pH and ionic strength [17,19,20]. Exceptions are hyaluronate (which lacks sulfate groups) and keratan sulfate (which lacks carboxyl groups) [19]. Those GAG with higher charge density or size bind to collagen with greater affinity, and the presence of L-iduronic residues appears to promote such binding.

PLASMA FIBRONECTIN STRUCTURE AND FUNCTIONAL DOMAINS

Our present knowledge of fibronectin structure and its functional domains reflects the contributions of many laboratories. Except for the summary to follow, mainly with GAG binding sites in the molecule, details and background concerning the current basis for

Table 1 General Features of Glycosaminoglycans

Type of polysaccharide	Repeating saccharide units	Sulfated	Occurrence in mammalian tissues
Hyaluronate	D-glucuronic D-glucosamine	No	Skin, vitreous humor, synovial fluid, umbilical cord, cartilage
Chondroitin 4- and 6-sulfate	D-glucuronic D-galactosamine	Yes	Cartilage, cornea, bone, skin, arterial wall
Dermatan sulfate	D-glucuronic L-iduronic	Yes	Skin, heart valves, tendon, arterial wall
Heparan sulfate	D-glucuronic L-iduronic D-glucosamine	Yes	Lung, arterial wall, cell surfaces
Heparin	D-glucuronic L-iduronic D-glucosamine	Yes	Lung, liver, skin, mast cells
Keratan sulfate	D-galactose D-glucosamine	Yes	Cartilage, cornea, intervertebral disks

its primary structure and for the various functional and domain assignments have been covered in several recent articles [11,13,16, 21-24]. Less recent articles should be consulted to provide complementary information on the development of knowledge of the structure and functions of this protein [25-27].

Most plasma fibronectin molecules are dimeric, have a molecular weight of ~450,000, and are composed of two disulfide-bridged chains of approximately equal size [28,29] termed A (larger) and B (smaller), respectively (Fig. 1). The size differences between these subunits are distinguishable by sodium dodecyl sulfate gel electrophoresis [4,28,30-33] and are accountable for by differences in the primary sequence of their COOH-terminal regions [4,34-38].

At least two separate domains in plasma fibronectin molecules bind to heparin (and to certain other competing GAG) under physiological pH and ionic strength conditions. The weaker (heparin I) is located in the NH_2-terminal region, whereas the most strongly binding domain (heparin II) is located in the COOH-terminal portion of each chain [10,14,16,33,34,36–40].

The binding behavior of the high-affinity heparin II site is insensitive to the presence (or absence) of Ca^{2+} [5,41], whereas the binding of heparin at the NH_2-terminal site is inhibited by Ca^{2+} but not by Mg^{2+} or Mn^{2+} [5]. Consistent with the electrostatic nature of heparin-fibronectin binding, the COOH-terminal regions are relatively basic, although there is an uneven distribution of charge clusters as reflected by the wide variation in the isoelectric points of cathespin D and plasmic peptide fragments obtained from this region of the molecule [36]. Despite that the human A and B chains do not differ from one another functionally with respect to heparin, cell, and fibrin binding features, the smaller B chain lacks a cathespin D- and plasmin-sensitive site present in the A chains and located in the region between the Heparin II and fibrin II binding sites [36,38].

Not all species' plasma fibronectins have the same functional binding domains, however. Isemura et al. [39] found that porcine plasma fibronectin had a less strongly binding heparin II site on its B chain compared with that on its A chain. Sekiguichi and coworkers [4,33] reported that the hamster plasma fibronectin B chain lacked a fibrin binding site such as that found in the human B chain.

Gold et al. [14] analyzed subtilisin-induced fragments of human plasma fibronectin, and in addition to confirming the location of the heparin I and II binding domains, they located a third, weakly binding heparin domain situated between the collagen and cell binding domains. This site is reportedly near or even identical with a DNA binding site stuided by Pande and Shively [15]. Hayashi and Yamada [5] also characterized a third heparin binding site inhibited by Ca^{2+}, Mg^{2+}, or Mn^{2+}. Its relationship to that described by Gold et al. [14] is not yet clear, although the latter authors believe that the sites are distinct. The existence on fibronectin molecules of several different heparin binding domains, some of which can be modulated by divalent cations, may indicate that each has a distinctive biological function.

Hyaluronate Binding to Fibronectin

Studies by Yamada et al. [41] suggested that chicken embryo fibroblast fibronectin had an affinity for hyaluronic acid as well as for heparin. The binding sites for each of these GAG were apparently separate from one another. In contrast, Laterra et al. [42,43] and Stamatoglou and Keller [44] showed that plasma fibronectin had no significant affinity for hyaluronic acid under any circumstance, although aggregated cellular forms of fibronectin did not bind to hyaluronate [43].Somewhat unexpectedly, Isemura et al. [45] found that porcine plasma fibronectin bound to hyaluronic acid-Sepharose 4B even more strongly than it did to heparin-Sepharose 4B. Furthermore, all major cathepsin D fragments from the porcine molecule bound to the hyaluronate column. However, like the findings on chicken fibroblast fibronectin [41], their results indicated that the heparin and hyaluronate binding sites on the porcine molecule were not the same.

HEPARIN- AND COLD-INDUCED PRECIPITATION OF PLASMA FIBRONECTIN

Thomas et al. [46] first made the observation that heparin could induce precipitation of plasma proteins in the cold. They reported the formation of a "heparin-precipitable fraction" (HPF) from the plasma of endotoxin-treated rabbits. Later studies by Smith and coworkers [47,48] indicated that a HPF could be produced in all normal plasmas, although larger amounts were found in plasmas from patients with inflammatory, infectious, and neoplastic diseases.

Smith and von Korff [47] also determined, on the basis of its coagulability as well as its electrophoretic and ultracentrifugal behavior, that fibrinogen was the major component of the HPF and that maximal formation of HPF depended upon the concentration of heparin and the presence of divalent cations. Precipitation of plasma was adversely affected by increasing the temperature, pH, or ionic strength. They also observed a second "minor" component similar to a protein in plasma subfraction I-1 [49]. That component, which now is known as plasma fibronectin, was then termed "cold-insoluble globulin." In addition, they found that other sulfated polysaccharides such as polymannuronic acid and chitin sulfate, also produced a cold-insoluble precipitate, whereas chondroitin sulfate induced only a small precipitate.

More recent studies have examined the specific role that plasma fibronectin plays in HPF formation [50,51]. Under physiological solvent conditions, heparin forms a cold-precipitable complex with purified plasma fibronectin or with mixtures of plasma fibronectin and fibrinogen but not with fibrinogen alone [50]. Consistent with earlier findings [47], precipitation can be augmented by addition of Ca^{2+} or by selecting optimal heparin levels; it can be abolished by raising the ionic strength or by raising the heparin concentration above the level required for maximum plasma fibronectin precipitation [50]. The presence of fibrinogen reduced the threshold for heparin-induced plasma fibronectin precipitation and, by coprecipitating with heparin and plasma fibronectin, increased the amount of precipitate. Thus, the heparin-plasma fibronectin interaction is crucial for formation of the HPF. Fibrinogen is not essential for precipitation but participates in HPF formation by interacting with plasma fibronectin via its Aα chains [50]. The observations that heparins with varying anticoagulant potential, as well as sulfated GAG with no anticoagulant activity, can produce a HPF [47,50,52] indicate that precipitate induction has little to do with heparin's anticoagulant potential.

Stathakis and Mosesson [50] suggested that little, if any, complex formation occurred between plasma fibronectin and heparin at 37°C, whereas at 22°C soluble complexes did form. On the other hand, although plasma fibronectin does not interact or precipitate with fibrinogen at temperatures between 37 and 2°C, fibrinogen solutions containing fibrin do form cold-insoluble complexes with plasma fibronectin [53]. Furthermore, fibrin has been shown to have a higher affinity for plasma fibronectin than does fibrinogen [53,54]. Jilek and Hörmann [52] reported that heparin enhances the precipitation of plasma fibronectin and native or denatured (gelatin) collagen at an ionic strength of 0.2 M at 4°C; hyaluronic acid and putrescine inhibited the heparin-fibronectin-induced precipitation of collagen, whereas chondroitin sulfate and dermatan sulfate had little effect. Other laboratories [55,56] have demonstrated that complex formation between fibronectin and heparin or other GAG (such as dermatan sulfate) enhances the interaction with collagen. Thus, formation of heparin-fibronectin complexes to which collagen or fibrinogen are added produces greater precipitation than in the absence of either macromolecule.

These above observations may be a reflection of low-temperature effects on plasma fibronectin itself. Mortillaro et al. [57] reported that the far-ultraviolet circular dichroism spectrum of plasmas fibronectin underwent changes between 37 and 5°C, suggesting that temperature-dependent structural changes occurred. More definitive studies are needed to ascertain the exact modulations that take place in plasma fibronectin due to temperature and/or heparin binding.

The plasma precipitate induced in the cold by heparin has a potentially practical value in that separation of certain functional components of the plasma factor VIII complex (that is, von Willebrand's Antigen, vWAg; ristocetin cofactor activity, VIIIR:RC; factor VIII procoagulant activity, VIII:C) occurs [58]. The larger vWAg multimers tend to precipitate in the HPF along with most of the associated VIIIR:RC activity. In contrast, the VIII:C activity remains exclusively in the supernatant plasma and can subsequently be concentrated by the standard cyroprecipitation technique of Pool et al. [59]. The ability to separate these biological activities by such simple methods may prove to be of value in the preparation of plasma concentrates for clinical use in treating hemophilia, von Willebrand's disease, and other conditions requiring replacement with plasma fibronectin.

GAG AND FIBRONECTIN AS COORDINATED EXTRACELLULAR MATRIX COMPONENTS

A number of observations indicate that GAG and cellular fibronectins in the extracellular matrix interact with each other as well as with collagen. Although it is clear that fibronectin and procollagens are codistributed in the pericellular connective tissue matrix [60,61], the relationship between these components and other matrix components (such as GAG) with which both fibronectin [50,55] and collagen [17] interact has been studied in some detail. Ruoslahti and Engvall [55] found that heparin enhanced the interaction between gelatin and fibronectin, as did Jilek and Hörmann [52], who showed that the addition of heparin caused precipitation of solutions of collagen and fibronectin. In addition, Johansson and Höök [56] showed that the rate and the

extent of binding of plasma fibronectin to both native and denatured collagen was enhanced by the presence of heparin or
highly sulfated dextran. Other GAG, however, including hyaluronic
acid, dermatan sulfate, chondroitin sulfate, or heparan sulfate,
were inactive.

Plasma fibronectin labeled with photoactivatable aryl azide
groups was added to normal and transformed fibroblasts; analyses
after activation by ultraviolet light revealed that the fibronectin
had been cross-linked to cell surface sulfated proteoglycans [62].
Yamada et al. [41] found that chick embryo fibroblast fibronectin bound with good affinity to hyaluronic acid and heparin (K_d,
10^{-7}-10^{-8}). This fibronectin bound more weakly to heparan sulfate and was minimally reactive with chondroitin sulfate and gly-
copeptides. The binding to hyaluronate and heparin was not
blocked by EDTA or by other GAG and was only moderately inhibited by elevated ionic strength.

Studies of the pericellular matrix produced by human embryonic skin fibroblasts [63] showed codistribution in the fibrillar
network among fibronectin, heparan sulfate proteoglycans, and a
portion of the chondroitin sulfate molecules. Associated analyses,
including affinity chromatography on immobilized fibronectin and
immunoprecipitation, indicated that the fibronectin-GAG bonds
were weaker than the bonds associated with the structural integrity of the matrix, thus suggesting that there were multiple molecular interactions stabilizing the fibronectin-procollagen matrix.
Similarly, Cossu and Warren [64] found that lactosaminoglycans
and heparan sulfate were covalently bound to fibronectin synthesized by undifferentiated embryonal carcinoma cells. Additional
support for a close interaction between GAG and fibronectin in
connective tissues is the observation that incorporation of heparin
into tissue extraction solutions increases the yield of fibronectin
extracted from lung parenchyma and placental villi [65].

It is not certain what role chondroitin sulfate plays in modulating cell attachment or in the formation of the fibronectin-collagen matrix, since it has affinity for collagen but does not bind to
fibronectin [41,66]. Only a fraction of the chondroitin sulfate
population was codistributed in the fibroblast fibrillar network
containing fibronectin and heparan sulfate [63]. Avian neural
crest cells migrated earlier and more extensively on a fibronectin-
coated surface; however, this migration was retarded by certain

proteoglycans, such as undersulfated chondroitin sulfate [67].
Similarly, Brennen et al. [68] showed that a chondroitin sulfate
PG from a rat yolk sac tumor diminished the adhesion of these
cells to fibronectin or collagen substrates. Cartilage proteoglycans
inhibited fibronectin-mediated fibroblast adhesion to collagen
[69,70], although chondroitin sulfate, which is the principal GAG
of cartilage (Table 1), had no significant effect on this process.

ROLE OF GAG IN CELL ATTACHMENT, SPREADING, AND LOCOMOTION

When a variety of cells adherent to the underlying substratum
are treated with the calcium chelator EGTA, they round up and
retract from these adhesion sites, leaving the material associated
with this site still firmly bound to the dish as "footprints" [71].
In the case of normal or transformed Balb/c 3T3 cells, newly
formed footprints are considerably enriched with heparan sulfate
and fibronectin [72,73]. Heparan sulfate and fibronectin, along
with cytoskeletal components coordinately resist solubilization
from substrate adhesion sites secondary to treatment with various
hydrolases [74] and guanidine hydrochloride [74,75]. Hyaluronate
and chondroitin PG are deposited as the murine fibroblasts begin
spreading over the substrate [42], but only a small portion of
radiolabeled fibronectin is bound to the substrate via these GAG.
Binding studies on GAG obtained from these cells indicated that
only highly sulfated heparan sulfate and a small subset of the dermatan sulfate population were able to bind to plasma fibronectin-
Sepharose affinity columns [42]. They could be displaced from
the column only by heparin, heparan sulfate, or dermatan sulfate.
    Studies of neural tumor cells (rat glioma or neuroblastoma)
also indicate that fibronectin-dependent adhesion to serum-coated
substrate may be mediated by heparan on their surface [76].
That heparan sulfate from nonadhering murine myeloma cells does
not bind plasma fibronectin [77], whereas that from adhering 3T3
cells does, emphasizes that not all components of a given type of
GAG display the same binding behavior. Such differences may be
a function of the degree of sulfation [42] or other less well defined factors.
    Attachment of several 3T3 cell lines to a heparin-binding sur-
face (platelet factor 4, PF4) was as effective as binding to a plasma

fibronectin surface. Binding to the PF4 substratum was blocked
by treating with heparin, or by treating the cells with heparinase
[78]. Heparinase treatment did not inhibit cell attachment to the
plasma fibronectin surface but did inhibit spreading. Thus, cell
surface heparan sulfate PG are critical mediators of cell adhesion,
and both heparan sulfate plus another cell component ("receptor")
are required for complete adhesion of fibroblasts [78,79]. They
act as mediators of cell adhesion, probably by binding to cell surfaces as well as to substrate-bound fibronectin. The additional ability of both fibronectin and GAG to bind to collagen would tend
to reinforce such coordinated interactions whenever this component is present. Klebe and Mock [80] also showed that the GAG
binding site of fibronectin plays a role in the mechanism of cell
attachment, in addition to the role played by its well-characterized
collagen binding site [81,82].

In contrast to the ability of hyaluronate to bind to cell surface
forms of fibronectin [41,43], the failure of human plasma fibronectin to bind hyaluronate indicates a significant difference in
both function and structure from the fibroblast molecule and may
explain why hyaluronate does not mediate adhesion of certain cell
lines to plasma fibronectin-coated culture dishes [78]. On the
other hand, hyaluronate, by virtue of its differential interactions
with aggregated versus nonaggregated forms of cellular fibronectin,
could play an important role in either stabilization or destabiliza-
tion of a cellular adhesion site, actions that could in turn function
as determinants of cellular adhesiveness and motility [43,67,83].

Kleinman et al. [84] demonstrated that preincubation of a
fibronectin-collagen matrix with gangliosides inhibited the attachment of Chinese hamster ovary (CHO) cells, although these sialic
acid-rich glycolipids did not inhibit the binding of fibronectin to
collagen. On this basis, they proposed a role for gangliosides as a
cell surface receptor for fibronectin. Yamada et al. [85] examined
this question further and found that gangliosides inhibited cellular
fibronectin-mediated hemagglutination, baby hamster kidney
(BHK) cell spreading, and restoration of normal morphology to
transformed cells. The inhibition was dose dependent and related
to the number of sialic acid residues in the oligosaccharide portions of the ganglioside. More recent studies have shown that retention of fibronectin at the cell surface was defective in a ganglioside-deficient cell line and that addition of gangliosides restored

a more normal fibrillar arrangement of fibronectin attached to the cell surface [86]. Perkins et al. [87] showed that gangliosides inhibited the spreading of both CHO and BHK cells on fibronectin-coated substrates and that they could bind directly to fibronectin with a relatively low affinity. Spreading of BALB/c 3T3 cells on a ganglioside-specific ligand-coated surface was less extensive than on a fibronectin-coated surface. Taken together, these findings suggest that the fibronectin "receptor" on certain cell types may contain a negatively charged glycolipid; however, the observed interactions between fibronectin and cell surface gangliosides seem inconsistent with the idea that gangliosides alone can act as the receptor for fibronectin.

EFFECTS OF GAG ON FIBRONECTIN-MEDIATED MONOCYTE AND MACROPHAGE FUNCTION

Monocytes and macrophages are phagocytic cells that collectively comprise the so-called mononuclear phagocytic system (also known as the reticuloendothelial system, RES). These cells perform critical functions in inflammation and in the immune response. Certain plasma proteins or their cleavage fragments can regulate mononuclear phagocytic cell function via recognition by plasma membrane receptors; such recognition leads to macrophage accumulation at an injury site followed by clearance of debris before and during tissue reconstruction [88]. In view of the widely studied interactions between plasma fibronectin and monocytes and macrophages, it is appropriate to consider the role of GAG in such a system. Blumenstock et al. [89] demonstrated the identity between plasma fibronectin and a serum protein they termed α_2-opsonic glycoprotein. This protein previously had been studied as a plasma factor promoting heparin-dependent uptake of a gelatin-coated emulsion by rat liver slices. This assay has been used as a measure of fibronectin-mediated phagocytosis by resident liver macrophages [90,91]. That uptake in their system is dependent upon the presence of heparin suggested a role for this GAG in the macrophage-fibronectin interaction.

To date, many investigators have examined the role of heparin in affecting the binding of fibronectin-coated particles, and it is clear that not all systems are dependent on heparin. Follow-up studies [92] have indicated that, in a peritoneal macrophage

monolayer system, heparin was not absolutely necessary for fibronectin-mediated particle uptake, although it did amplify the response. Molnar et al. [93] reported that a component of commercial heparin with no anticoagulant activity was required for promoting the uptake of ^{125}I-labeled gelatin-coated latex particles by
rat liver slices.

To examine directly the question of macrophage function,
macrophage binding assays have been developed to supplement the
type of indirect results that accrued from the liver slice assay system. Different conclusions have been drawn concerning the putative role of heparin in mediating binding and/or uptake by such
cells. Jilek and Hörmann [94] reported that plasma fibronectin
mediated the binding of fibrin monomer, as well as that of denatured collagen, to suspensions of peritoneal macrophages. These
reactions occurred in the absence of added heparin. Others, using
monocyte-macrophage monolayer systems, found no enhancing
activity for heparin in their systems [95-97]. In other experiments
representing a "middle ground", heparin was found to be unneces-
sary for fibronectin-mediated cell binding and uptake of gelatin-
coated particles, although its presence did augment this activity
[93,98-103].

Other GAG, such as heparin sulfate, have also been shown to
enhance binding to macrophages, whereas chondroitin sulfates A
and C, dermatan sulfate, and keratan sulfate did not [101,102]. In
contrast to the augmentation of fibronectin binding to gelatin-
coated particles by heparin, fibronectin-mediated binding of *Staphylococcus aureus* to macrophages was independent of the presence of heparin [104,105].

Some of these studies suggest that GAG may play a role as a
cofactor in phagocytosis of fibrin-fibrinogen complexes and/or denatured collagen during the wound-healing process. However, this
function may operate by mechanisms that do not directly affect
the cells themselves: (1) GAG like heparin may influence macrophages by enhancing the binding of fibronectin to native or denatured collagen [44,55,56,106]; (2) heparin and other GAG may
increase the binding of gelatin partcles to macrophages by enhancing their agglutination to one another, thus leading to amplifica-
tion of the number of particles becoming cell bound [55,56,102,
106].

CONCLUDING REMARKS

In this chapter, we have summarized current knowledge of interactions between glycosaminoglycans, proteoglycans, fibronectin, and other elements of the blood and connective tissue. For one thing, interactions between GAG and plasma fibronectin result in their coprecipitation in the cold, along with fibrinogen and certain components of the factor VIII-von Willebrand complex. This phenomenon may provide a useful means for preparing clinically valuable concentrates of plasma proteins. More importantly, these macromolecules are key structural components of the extracellular and pericellular matrix. Together, they participate in the vital biological processes of cell attachment, spreading, and locomotion and perhaps as well in the phagocytic functions of monocytes and macrophages. Furthering our understanding of the influence of these macromolecules on such biological processes will continue to be an important subject of investigation for some time to come.

ACKNOWLEDGMENTS

The work carried out in the authors' laboratory is supported by an NIH program project grant HL-28444, and NIH New Investigator Award (to David L. Amrani, grant R23-AM32762-01), and by grants-in-aid from the Wisconsin Affiliate of the American Heart Association. We gratefully acknowledge Ms. Julie Erickson's contributions in organizing and typing the manuscript.

REFERENCES

1. J. Keski-Oja and K. M. Yamada, Isolation of an actin-binding fragment of fibronectin. *Biochem. J. 193*:615-620 (1981).
2. D. F. Mosher and R. A. Proctor, Binding and factor XIIIa-mediated cross-linking of a 27-kilodalton fragment of fibronectin to staphylococcus aureus. *Science 209*:297-929 (1980).
3. H. Hörmann and M. Seidl, Affinity chromatography on immobilized fibrin monomer III. The fibrin affinity center of fibronectin. *Hoppe-Seyler's Z. Physiol. Chem. 361*:1449-1452 (1980).
4. K. Sekiguchi, M. Fukuda, and S. Hakomori, Domain structure of hamster plasma fibronectin. Isolation and characterization of four functionally distinct domains and their unequal distribution between two subunit polypeptides. *J. Biol. Chem. 256*:6452-6462 (1981).

5. M. Hayashi and K. M. Yamada, Divalent cation modulation of fibronectin binding to heparin and to DNA. *J. Biol. Chem.* 257:5263-5267 (1982).
6. G. Balian, E. M. Click, E. Crouch, J. M. Davidson, and P. Bornstein, Isolation of a collagen-binding fragment from fibronectin and CIg. *J. Biol. Chem.* 254:1429-1432 (1979).
7. G. Balian, E. M. Click, and P. Bornstein, Location of a collagen-binding domain in fibronectin. *J. Biol. Chem.* 255:3234-3236 (1980).
8. M. B. Furie and D. B. Rifkin, Proteolytically derived fragments of human plasma fibronectin and their localization within the intact molecule. *J. Biol. Chem.* 255:3134-3140 (1980).
9. M. B. Furie, A. B. Frey, and D. B. Rifkin, Location of a gelatin-binding region of human plasma fibronectin. *J. Biol. Chem.* 255:4391-4394 (1980).
10. E. Ruoslahti, E. G. Hayman, E. Engvall, W. C. Cothran, and W. T. Butler, Alignment of biologically active domains in the fibronectin molecule. *J. Biol. Chem.* 256:7277-7281 (1981).
11. T. E. Petersen, H. C. Thøgersen, K. Skorstengaard, K. Vibe-Pedersen, P. Sahl, L. Sottrup-Jensen, and S. Magnusson, Partial primary structure of bovine plasma fibronectin; three types of internal homology. *Proc. Natl. Acad. Sci. USA* 80:137-144 (1983).
12. R. Ehrismann, M. Chiquet, and D. C. Turner, Mode of action of fibronectin in promoting chicken myoblast attachment. Mr = 60,000 gelatin-binding fragment binds native fibronectin. *J. Biol. Chem.* 256:4056-4062 (1981).
13. K. M. Yamada, Cell surface interactions with extracellular materials. *Annu. Rev. Biochem.* 52:761-799 (1983).
14. L. I. Gold, B. Frangione, and E. Pearlstein, Biochemical and immunological characterization of three binding sites on human plasma fibronectin with different affinities for heparin. *Biochemistry* 22:4113-4118 (1983).
15. H. Pande and J. E. Shively, NH_2-terminal sequences of DNA-, heparin-, and gelatin-binding tryptic fragments from human plasma fibronectin. *Arch. Biochem. Biophys.* 213:258-265 (1982).
16. R. Ehrismann, D. E. Roth, H. M. Eppenberger, and D. C. Turner, Arrangement of attachment-promoting, self-association, and heparin-binding sites in horse serum fibronectin. *J. Biol. Chem.* 257:7381-7387 (1982).
17. V. Lindahl and M. Höök, Glycosaminoglycans and their binding to biological macromolecules. *Annu. Rev. Biochem.* 47:385-417 (1978).
18. V. C. Hascall and G. K. Hascall, Proteoglycans, In *Cell Biology of Extracellular Matrix* (E. D. Hay, Ed.). Plenum Press, New York, 1981, pp. 39-63.
19. R. A. Greenwald, C. E. Schwartz, and J. O. Cantor, Interaction of cartilage proteoglycans with collagen-substituted agarose gels. *Biochem. J.* 145:601-605 (1975).

92 MOSESSON AND AMRANI

20. T. R. Oegema, J. Laidlow, V. C. Hascall, and D. D. Dziewiatkowski, The effect of proteoglycans on the formation of fibrils from collagen solutions. *Arch. Biochem. Biophys. 170*:698-709 (1975).
21. S. K. Akiyama, K. M. Yamada, and M. Hayaski, The structure of fibronectin and its role in cellular adhesion. *J. Supramol. Struct. Cell. Biochem. 16*:345-358 (1981).
22. H. Hörmann, Fibronectin—mediator between cells and connective tissue. *Klin. Wochenschr. 60*:1265-1277 (1982).
23. R. O. Hynes and K. M. Yamada, Fibronectins: Multifunctional modular glycoproteins. *J. Cell. Biol. 95*:369-377 (1982).
24. L. T. Furcht, Structure and function of the adhesive glycoprotein fibronectin. *Mod. Cell Biol. 1*:53-117 (1983).
25. K. M. Yamada and K. Olden, Fibronectins—adhesive glycoproteins of cell surface and blood. *Nature 275*:179-184 (1978).
26. M. W. Mosesson and D. L. Amrani, The structure and biologic activities of plasma fibronectin. *Blood 56*:145-158 (1980).
27. D. F. Mosher, Fibronectin. *Prog. Hemost. Thromb. 5*:111-151 (1980).
28. M. W. Mosesson, A. B. Chen, and R. M. Huseby, The cold-insoluble globulin of human plasma: Studies of its essential structural features. *Biochim. Biophys. Acta 386*:509-524 (1975).
29. D. F. Mosher, Crosslinking of cold-insoluble globulin by fibrin-stabilizing factor. *J. Biol. Chem. 250*:6614-6621 (1975).
30. A. B. Chen, D. L. Amrani, and M. W. Mosesson, Heterogeneity of the cold-insoluble globulin of human plasma. *Biochim. Biophys. Acta 493*: 310-322 (1977).
31. F. Jilek and H. Hörmann, Cold-insoluble globulin: Plasminolysis of cold-insoluble globulin. *Hoppe-Seyler's Z. Physiol. Chem. 358*:133-136 (1977).
32. M. Kurkinen, T. Vartio, and A. Vaheri, Polypeptides of human plasma fibronectin are similar but not identical. *Biochim. Biophys. Acta 624*: 490-498 (1980).
33. K. Sekiguchi and S. Hakomori, Domain structure of human plasma fibronectin. Differences and similarities between human and hamster fibronectins. *J. Biol. Chem. 258*:3967-3973 (1983).
34. H. Richter, M. Seidl, and H. Hörmann, Location of heparin-binding sites of fibronectin. Detection of a hitherto unrecognized transamidase sensitive site. *Hoppe-Seyler's Z. Physiol. Chem. 362*:399-408 (1981).
35. B. T. Atherton and R. O. Hynes, A difference between plasma and cellular fibronectins located with monoclonal antibodies. *Cell. 25*:133-141 (1981).
36. H. Richter and H. Hörmann, Early and late cathepsin D-derived fragments of fibronectin containing the C-terminal interchain disulfide crosslink. *Hoppe-Seylers Z. Physiol. Chem. 363*:351-364 (1982).

37. D. E. Smith and L. T. Furcht, Localization of two unique heparin binding domains of human plasma fibronectin with monoclonal antibodies. *J. Biol. Chem. 257*:6518-6523 (1982).
38. M. Hayashi and K. M. Yamada, Domain structure of the carboxyl-terminal half of human plasma fibronectin. *J. Biol. Chem. 258*:3332-3340 (1983).
39. M. Isemura, Z. Yosizawa, K. Takahashi, H. Kasaka, N. Kojima, and T. Ono, Characterization of porcine plasma fibronectin and its fragmentation by porcine liver cathepsin B. *J. Biochem. 90*:1-9 (1981).
40. T. Vartio, Characterization of the binding domains in the fragments cleaved by cathepsin G from human plasma fibronectin. *Eur. J. Biochem. 123*:223-233 (1982).
41. K. M. Yamada, D. W. Kennedy, K. Kimata, and R. M. Pratt, Characteristics of fibronectin interactions with glycosaminoglycans and identification of active proteolytic fragments. *J. Biol. Chem. 255*:6055-6063 (1980).
42. J. Laterra, R. Ansbacher, and L. A. Culp, Glycosaminoglycans that bind cold-insoluble globulin in cell-substratum fibroblasts. *Proc. Natl. Acad. Sci. USA 77*:6662-6666 (1980).
43. J. Laterra and L. A. Culp, Differences in hyaluronate binding to plasma and cell surface fibronectins. *J. Biol. Chem. 257*:719-726 (1982).
44. S. C. Stamatoglou and J. M. Keller, Interactions of cellular glycosaminoglycans with plasma fibronectin and collagen. *Biochim. Biophys. Acta 719*:90-97 (1982).
45. M. Isemura, Z. Yosizawa, T. Koide, and T. Ono, Interaction of fibronectin and its proteolytic fragments. *J. Biochem. 91*:731-734 (1982).
46. L. Thomas, R. T. Smith, and R. W. von Korff, Cold precipitation by heparin of a protein in rabbit and human plasma. *Proc. Soc. Exp. Biol. Med. 86*:813-819 (1954).
47. R. T. Smith and R. W. von Korff, A heparin-precipitable fraction of human plasma. I. Isolation and characterization of the fraction. *J. Clin. Invest. 36*:596-604 (1957).
48. R. T. Smith, A heparin-precipitable fraction of human plasma II. Occurrence and significance of the fraction in normal individuals and in various disease states. *J. Clin. Invest. 36*:605-616 (1957).
49. P. R. Morrison, J. T. Edsall, and S. G. Miller, Preparation and properties of serum and plasma proteins. XVIII. The separation of purified fibrinogen from fraction I of human plasma. *J. Am. Chem. Soc. 70*:3103-3109 (1948).
50. N. E. Stathakis and M. W. Mosesson, Interactions among heparin, CIg, and fibrinogen in formation of the heparin-precipitable fraction of plasma. *J. Clin. Invest. 60*:855-865 (1977).

51. O. Fyrand and N. D. Solum, Heparin precipitable fraction (HPF) from dermatological patients. Studies on the nonclottable components. Identification of cold-insoluble globulin as the major non-clottable component. *Thromb. Res. 8*:659-672 (1976).

52. F. Jilek and H. Hörmann, Fibronectin (cold-insoluble globulin). VI. Influence of heparin and hyaluronic acid on the binding of native collagen. *Hoppe-Seylers Z. Physiol. Chem. 360*:597-603 (1979).

53. N. E. Stathakis, M. W. Mosesson, A. B. Chen, and D. K. Galanakis, Cryoprecipitation of fibrin-fibrinogen complexes induced by the cold-insoluble globulin of plasma. *Blood 51*:1211-1222 (1978).

54. E. Engvall, E. Ruoslahti, and E. J. Miller, Affinity of fibronectin to collagens of different genetic types and to fibrinogen. *J. Exp. Med. 147*: 1584-1595 (1978).

55. E. Ruoslahti and E. Engvall, Complexing of fibronectin, glycosaminoglycans and collagen. *Biochim. Biophys. Acta 631*:350-358 (1980).

56. S. Johansson and M. Höök, Heparin enhances the rate of binding of fibronectin to collagen. *Biochem. J. 187*:521-524 (1980).

57. J. M. Mortillaro, D. L. Amrani, M. W. Mosesson, and N. M. Tooney, Low temperature effects on plasma fibronectin. *Biophys. J. 37*:250a (1982).

58. D. L. Amrani, M. W. Mosesson, and L. W. Hoyer, Distribution of plasma fibronectin (cold-insoluble globulin) and components of the factor VIII complex after heparin-induced precipitation of plasma. *Blood 159*:657-663 (1982).

59. J. G. Pool, E. J. Hershgold, and A. R. Pappenhagen, High potency antihemophiliac factor concentrate prepared from cryoglobulin precipitate. *Nature 203*:312 (1964).

60. P. Bornstein and J. F. Ash, Cell-surface associated structural proteins in connective tissue cells. *Proc. Natl. Acad. Sci. USA 74*:2480-2484 (1977).

61. A. Vaheri, M. Kurkinen, V. P. Lehto, E. Linder, and R. Timpl, Codistribution of pericellular matrix proteins in cultured fibroblasts and loss in transformation: Fibronectin and procollagen. *Proc. Natl. Acad. Sci. USA 75*:4944-4948 (1978).

62. M. E. Perkins, T. H. Ji, and R. O. Hynes, Crosslinking of fibronectin to sulfated proteoglycans at the cell surface. *Cell 16*:941-952 (1979).

63. K. Hedman, S. Johansson, T. Vartio, L. Kjellen, A. Vaheri, and M. Höök, Structure of the pericellular matrix: Association of heparan and chondroitin sulfates with fibronectin-procollagen fibers. *Cell 28*:663-671 (1982).

64. G. Cossu and L. Warren, Lactosaminoglycans and heparan sulfate are covalently bound to fibronectins synthesized by mouse stem teratocarcinoma cells. *J. Biol. Chem. 258*:5603-5607 (1983).

65. B. A. Bray, and G. M. Turino, Heparin facilitates the extraction of tissue fibronectin. *Science 214*:793-795 (1981).

66. A. Oldberg and E. Ruoslahti, Interactions between chrondroitin sulfate, proteoglycan, fibronectin, and collagen. *J. Biol. Chem. 257*:4859-4863 (1982).

67. D. F. Newgreen, I. L. Gibbins, J. Sauter, B. Wallenfels, and R. Wutz, Ultrastructural and tissue-culture studies in the role of fibronectins, collagen, glycosaminoglycans in the migration of neural crest cells in the fowl embryo. *Cell Tissue Res. 221*:521-549 (1982).

68. M. J. Brennan, A. Oldberg, E. G. Hayman, and E. Ruoslahti, Effect of a proteoglycan produced by rat tumor cells on their adhestion to fibronectin-collagen substrata. *Cancer Res. 43*:4302-4307 (1983).

69. P. Knox and P. Wells, Cell adhesion and proteoglycans. I. The effect of exogenous proteoglycans on the attachment of chick embryo fibroblasts to tissue culture plastic and collagen. *J. Cell Sci. 40*:77-88 (1979).

70. A. M. Rich, E. Pearlstein, G. Weissmann, and S. T. Hoffstein, Cartilage proteoglycans inhibit fibronectin-mediated adhesion. *Nature 293*: 224-226 (1981).

71. I. J. Rosen and L. A. Culp, Morphology and cellular origins of substrate attached material from mouse fibroblasts. *Exp. Cell Res. 107*:139-149 (1977).

72. L. A. Culp, B. A. Murray, and B. J. Rollins, Fibronectin and proteoglycans as determinants of cell-substratum adhesion. *J. Sipramol. Struct. 11*: 401-427 (1979).

73. B. A. Murray and L. A. Culp, Multiple and masked pools of fibronectin in murine fibroblast cell-substratum adhesion sites. *Exp. Cell. Res. 131*: 237-249 (1981).

74. L. A. Culp, P. J. Rollins, J. Buniel, and S. Hitri, Two functionally distinct pools of glycosaminoglycans in the substrate adhesion site of murine cells. *J. Cell Biol. 79*:788-801 (1978).

75. B. J. Rollins and L. A. Culp, Prelimary characterization of the proteoglycans in the substrate adhesion sites of normal murine cells. *Biochemistry 18*:141-148 (1979).

76. L. A. Culp and C. Domen, Plasma fibronectin-binding glycosaminoglycans in the substratum adhesion sites of neural tumor cells. *Arch. Biochem. Biophys. 213*:726-730 (1982).

77. S. C. Stamatoglou and J. M. Keller, Correlation between cell substrate attachment in vitro and cell surface heparin sulfate affinity for fibronectin and collagen. *J. Cell Biol. 96*:1820-1823 (1983).

78. J. Laterra, J. E. Silbert, and L. A. Culp, Cell surface heparan sulfate mediates some adhesive responses to glycosaminoglycan-binding matrices, including fibronectin. *J. Cell Biol. 96*:112-123 (1983).

79. J. Laterra, E. K. Norton, C. S. Izzard, and L. A. Culp, Contact formation by fibroblasts adhering to heparan sulfate binding substrate. *Exp. Cell Res. 146*:15-27 (1983).

80. R. J. Klebe and P. J. Mock, Effect of glycosaminoglycans on fibronectin-mediated cell attachment. *J. Cell Physiol. 112*:5-9 (1982).
81. R. J. Klebe, Isolation of a collagen-dependent cell attachment factor. *Nature 250*:248-251 (1974).
82. E. Pearlstein, Substrate activation of cell adhesion factor as a prerequisite for cell attachment. *Int. J. Cancer 22*:32-35 (1977).
83. B. J. Barnhart, S. H. Cox, and P. M. Kraemer, Detachment variants of chinese hamster cells: Hyaluronic acid as a modulator of cell detachment. *Exp. Cell Res. 119*:327-332 (1979).
84. H. K. Kleinman, G. R. Martin, and P. H. Fishman, Ganglioside inhibition of fibronectin-mediated cell adhesion to collagen. *Proc. Natl. Acad. Sci. USA 76*:3367-3371 (1979).
85. K. M. Yamada, D. W. Kennedy, G. R. Grotendorst, and T. Momoi, Glycolipids: Receptors for fibronectin? *J. Cell. Physiol. 109*:343-351 (1981).
86. K. M. Yamada, D. R. Critchley, P. H. Fishman, and J. Moss, Exogenous gangliosides enhance the interaction of fibronectin with ganglioside-deficient cells. *Exp. Cell Res. 143*:295-302 (1983).
87. R. M. Perkins, S. Keelie, B. Patel, and D. R. Critchley, Gangliosides as receptors for fibronectin? Comparison of cell spreading on a ganglioside-specific ligand with that on fibronectin. *Exp. Cell Res. 141*:231-243 (1982).
88. C. Bianco, O. Götze, and Z. A. Cohn, Complement, coagulation and mononuclear phagocytes, in *Mononuclear Phagocytes* (R. van Furth, ed.). Martinus Nijhoff, the Hague, 1980, pp. 1443-1458.
89. F. A. Blumenstock, T. M. Saba, P. Weber, and R. Laffin, Biochemical and immunological characterization of hyman opsonic alpha-2-SB glycoprotein: Its identity with cold-insoluble globulin. *J. Biol. Chem. 253*:4287-4291 (1978).
90. F. A. Blumenstock, T. M. Saba, P. B. Weber, and E. Cho, Purification and biochemical characterization of a macrophage stimulating alpha-2-globulin opsonic protein. *J. Reticuleondothel. Soc. 19*:157-172 (1976).
91. T. M. Saba, F. A. Blumenstock, P. Weber, and J. E. Kaplan, Physiologic role of cold-insoluble globulin in systemic host defense: Implications of its characterization as the opsonic alpha 2 surface-binding glycoprotein. *Ann. N.Y. Acad. Sci. 312*:43-55 (1978).
92. F. A. Blumenstock, T. M. Saba, E. Roccario, E. Cho, and J. E. Kaplan, Opsonic fibronectin after trauma and particle injection determined by a peritoneal macrophage monolayer assay. *J. Reticuloendothel. Soc. 30*: 61-71 (1981).
93. J. Molnar, F. B. Gelder, H. Z. Lai, G. E. Siefring, Jr., R. B. Credo, and L. Lorand, Purification of opsonically active plasma and rat cold-insoluble globulin (plasma fibronectin). *Biochemistry 18*:3909-3916 (1979).

94. F. Jilek and H. Hörmann, Fibronectin (cold-insoluble globulin). V. Mediation of fibrin-monomer binding to macrophages. *Hoppe-Seylers Z. Physiol. Chem. 359*:1603-1605 (1978).

95. M. P. Bevilacqua, D. Amrani, M. W. Mosesson, and C. Bianco, Receptors for cold-insoluble globulin (plasma fibronectin) on human monocytes. *J. Exp. Med. 153*:42-60 (1981).

96. J. K. Czop, J. L. Kadish, and K. F. Austen, Augmentation of human monocyte opsonin-independent phagocytosis by fragments of human plasma fibronectin. *Proc. Natl. Acad. Sci. USA 78*:3649-3653 (1981).

97. C. G. Pommier, S. Inada, L. F. Fries, T. Takahashi, M. M. Frank, and E. J. Brown, Plasma fibronectin enahnces phagocytosis of opsonized particles by human peripheral blood monocytes. *J. Exp. Med. 157*: 1844-1854 (1983).

98. P. W. Gudewicz, J. Molnar, M. Z. Lai, D. W. Beezhold, G. E. Siefring, R. B. Credo, and L. Lorand, Fibronectin-mediated uptake of gelatin-coated latex particles by peritoneal macrophages. *J. Cell. Biol. 87*: 427-433 (1980).

99. J. E. Doran, A. R. Mansberger, and A. C. Reese, Cold insoluble globulin-enahnced phagocytosis of gelatinized targets by macrophage monolayers: A model system. *J. Reticuloendothel. Soc. 27*:471-483 (1980).

100. H. Hörmann and V. Jelinic, Heparin enhances the binding of ^{125}I-fibronectin by macrophages. *Hoppe-Seylers Z. Physiol. Chem. 34*:379-387 (1980).

101. H. Hörmann and V. Jelinic, Regulation by heparin and hyaluronic acid of the fibronectin-dependent association of collagen, Type III, with macrophages. *Hoppe-Seylers Z. Physiol. Chem. 362*:87-94 (1981).

102. L. Van de Water, S. Schroeder, E. B. Crenshaw, III, and R. O. Hynes, Phagocytosis of gelatin-latex particles by a murine macrophage line is dependent on fibronectin and heparin. *J. Cell Biol. 90*:32-39 (1981).

103. H. Rieder, M. Birmelin, and K. Decker, Synthesis and functions of fibronectin in rat liver cells in vitro, in *Sinusoidal Liver Cells* (D. L. Knook and E. Wisse, eds.). Elsevier Biomedical Press, Amsterdam, 1982, pp. 193-200.

104. H. A. Verbrugh, P. K. Peterson, D. E. Smith, B. Y. Nguyen, J. R. Hoidal, B. J. Wilkinson, J. Verhoef, and L. T. Furcht, Human fibronectin binding to staphylococcal surface protein and its relative inefficiency in promoting phagocytosis by human polymorphonuclear leukocytes, monocytes, and alveolar macrophages. *Infect. Immun. 33*:811-819 (1981).

105. M. E. Lanser and T. M. Saba, Fibronectin as a co-factor necessary for optimal granulocyte phagocytosis of *Staphylococcus aureus*. *J. Reticuloendothel. Soc. 30*:415-424 (1981).

106. I. J. Check, H. C. Wolfman, T. B. Coley, and R. L. Hunter, Agglutination assay for human opsonic factor using felatin-coated latex particles. *J. Reticuloendothel. Soc. 25*:351-352 (1979).

6

Interaction with Fibrinogen and Fibrin

Helmut Hörmann
Max Planck-Institut für Biochemie
Martinsried bei München
Federal Republic of Germany

Fibronectin is a high-molecular-weight protein present as a soluble molecule in plasma and as fibrils on the surface of adherent cells (for recent reviews see Refs. 1 to 4). It expresses affinity to various extracellular substrates, including collagen, heparin, and fibrin. In addition, it has a binding site for cell surface receptors [5] and thus is capable of mediating various interactions of cells with those compounds. It is important for attachment and spreading of cells on various extracellular structures and is involved in the locomotion of the cells along those substrates. Fibronectin is also necessary for binding of bacteria, effete proteins, and cellular waste products by cells of the reticuloendothelial system and thus becomes an essential factor of the defense and blood clearing system [6,7].

The interaction between fibronectin and fibrin modifies and retards the coagulation process [8, 9]. Furthermore, it may be important for the attachment of fibroblasts on a fibrin coagulum [10], which subsequently is contracted by the cells [11]. In a wound area, connective tissue cells can move along fibrin strands into the affected region and are not swept away by fluids. Finally, fibronectin is assumed to play a basic role in the removal of soluble fibrin from circulating blood by cells of the reticuloendothelial system. At least, one observes a decline of the fibronectin plasma level as a consequence of trauma with major coagulation events or of illness with severe disseminated intravascular coagulation [12].

The affinity between fibronectin and fibrin is relatively weak and is experimentally demonstrable predominantly in the cold. At body temperature only weak associations exist, which, however, still might be of physiological importance. Less pronounced is the affinity to fibrinogen, which is only incompletely bound by fibronectin even at low temperature.

NONCOVALENT AND COVALENT AGGREGATES

Fibronectin-Fibrinogen Complexes in Cryoprecipitates

In the course of fibrinogen isolation from Cohn fraction I of human plasma, Morrison et al. [13] in 1948 mentioned a nonclottable high-molecular-weight by-product of crude fibrinogen (fraction-I-0) which could be partially removed by precipitation in the cold. The cryoprecipitate (fraction I-1) contained fibrinogen and 25–30% of the component termed "cold-insoluble globulin" with an electrophoretic mobility of a β_1-globulin and a sedimentation constant of 12–13S. The electrophoretic behavior of fraction I-1 in the U-tube apparatus conventional at that time, as well as sedimentation studies, already indicated partial interaction of the two components in fraction I-1 [14].

The noncovalent association of fibrinogen and cold-insoluble globulin, which later was termed plasma fibronectin, is remarkably enhanced by heparin. In 1954, Thomas et al. [15] described the formation of cryoprecipitates in heparin-stabilized plasmas of rabbits treated with endotoxin. Plasma of patients suffering from various inflammatory, infectious, or neoplastic diseases also ex-

creted considerable precipitates in the cold [16, 17]. Plasma of
normal donors yielded this "heparin precipitable plasma fraction"
to a considerable less content [17, 18]. Smith and von Korff [16]
investigated the influence of temperature, ionic strength, pH, and
heparin concentration on the precipitation and observed augmentation in the presence of divalent cations. Since heparin induced
cryoprecipitates in plasma but not in serum, fibrinogen appeared
to be a main component, although purified fibrinogen yielded
little cryoprecipitate. A second component identical with that present in fibrinogen fraction I-1 appeared essential. In 1970, this
cold-insoluble globulin was isolated from plasma and shown to be
different from fibrinogen by various criteria [19].

Fyrand et al. [20] found about 60% coagulability of redis-
solved cryoprecipitates obtained from heparinized plasma of
dermatological patients. After removal of fibrinogen by heating at
56°C, the supernatant contained predominantly cold-insoluble globulin [21]. Similar observations were reported by Matsuda et al.
[22] on patients with skin ulcers on their legs.

Formation of cryoprecipitates in a system containing fibrinogen, heparin, and cold-insoluble globulin (plasma fibronectin) was
more closely investigated by Stathakis and Mosesson [23]. Heparin
induced precipitation of plasma fibronectin free of fibrinogen, and
addition of fibrinogen increased the mass of precipitate until an
upper limit was reached at about 300 μg/ml. An optimal heparin
concentration was found at 125 μg/ml, beyond which precipitation
declined. Precipitation was readily observed at ionic strength 0.1
and pH 7.2 and vanished at 0.2 or at higher pH values. Calcium
ions were found to improve the precipitation.

Fibrinogen potentiated the formation of precipitates from fi-
bronectin- and heparin-containing solutions in the cold, but it
failed to influence the shape of the precipitation curve with respect
to heparin concentration, suggesting that fibrinogen associated
with a fibronectin-containing nucleus. The two proteins were identified in the cryoprecipitates by gel electrophoresis. In addition,
heparin was present as revealed from the anticoagulant activity of
the precipitates.

Thus far it is not clear why heparin induces cryoprecipitates
more readily from plasma of patients with dermatological diseases
or ulcers than from normal plasma. Fibrinogen appears not to be
modified, as indicated by end-group analysis and gel electrophoresis

of the subunit peptide chains. In addition, the pathological plasmas showed a normal fibronectin content, and degradation could not be observed [22].

Fibronectin can be separated from fibrinogen by chromatography on gelatin-Sepharose at room temperature [24]. Fibronectin is selectively retained and can be desorbed with 1.0 M potassium bromide, pH 5.3 [25] or 1.0 M arginine [26]. At 4°C, however, fibronectin and fibrinogen are retained together by gelatin-Sepharose [27] (Table 1). Separation is also possible by chromatography on DEAE-cellulose [28].

A soluble covalent complex of fibronectin and fibrinogen was detected in plasma of patients suffering from acute leukemia [29]. The complex was termed FN:C.

Fibronectin-Fibrin Interaction

Interaction between fibronectin and fibrin can take place noncovalently by affinity forces as well as by covalent cross-linking. The affinity to fibrin is definitely stronger than that to fibrinogen. Complexes formed in the absence of heparin are stable in the cold and can be detected by cryoprecipitation or affinity chromatography at low temperatures. At 37°C, however, they dissociate and end most likely in association equilibria. Therefore, if binding is desired to be stable at body temperature, covalent cross-linking of the two proteins is necessary.

Noncovalent Binding

Binding of plasma fibronectin to fibrin was first described by Ruoslahti and Vaheri [30], although relevant observations were pub-

Table 1 Graduated Affinity of Fibrinogen and Fibronectin to Gelatin-Sepharose at pH 7.4, Ionic Strength 0.15

	Retention	
Protein	4°C	20°C
Fibrinogen	+	-
Fibronectin	+	+

lished previously [19]. A plasma protein immunologically identified as cold-insoluble globulin was absent in serum, if plasma was coagulated at $0°C$, but was present in large amounts in serum prepared at $37°C$ (see also Refs. 25 and 31).

If plasma was chromatographed on fibrin-Sepharose at $4°C$ [32], various fractions of fibrinogen and a high-molecular-weight substance later identified as cold-insoluble globulin (fibronectin) [33] were retained. This protein as well as early fibrinogen degradation products were eluted when the temperature was raised to $37°C$, but intact fibrinogen was not desorbed unless 1 M potassium bromide, pH 5.3, was applied [32]. Fibrinogen-Sepharose retained at $4°C$ fibronectin and fibrin monomer, if plasma previously subjected to a limited thrombin treatment was applied [33]. Since fibronectin in untreated plasma was not retained by fibrinogen-Sepharose, it was concluded that, in thrombin-treated plasma, it was bound to fibrin monomer retained by fibrinogen-Sepharose. The two proteins were eluted with 1 M potassium bromide, pH 5.3.

Using purified substances, Stathakis and coworkers [34] showed binding of plasma fibronectin to fibrin-Sepharose as well as to fibrinogen-Sepharose at $4°C$ and pH 7.6. Absorbed fibronectin was entirely eluted at $22°C$ from fibrinogen-Sepharose, whereas from fibrin-Sepharose only a fraction was obtained, with the rest eluted with 6 M urea in 0.5 M sodium chloride, pH 4.1 (see also Ref. 31).

The same authors [34] also showed that fibrin monomer, kept in solution by excess fibrinogen, gave rise to cryoprecipitates if fibronectin was present but remained soluble in its absence. Similarly, plasma in which soluble fibrin had been generated by limited thrombin treatment formed complexes of fibrin-fibrinogen and fibronectin in the cold. The fibronectin content of the precipitates was relatively low. Analysis [35] yielded a molecular ratio fibronectin-fibrinogen-fibrin of approximately 0.05:0.8:0.2. The authors therefore concluded that fibronectin might form a nucleus to which fibrin-fibrinogen complexes can attach. Kaplan and Snedeker [8] observed that fibronectin kept fibrin monomer at $37°C$ in solution and was even capable of dissolving a fibrin coagulum provided it was not stabilized. At a ratio of fibronectin to fibrin of 1:6, about 90% of the fibrin remained soluble at $37°C$. Niewiarowska and Cierniewski [9] also reported that fibronectin retarded coagulation of fibrin generated by thrombin or even prevented the process if present in high amounts.

Binding of fibronectin to a fibrin coagulum was inhibited by gelatin, as demonstrated by Engvall et al. [36]. An ELISA technique yielded considerably less affinity of fibronectin to fibrinogen than to gelatin.

Covalent Cross-linking of Fibronectin with Fibrin

Mosher [37, 38] detected that fibronectin was a substrate for thrombin-activated plasma transamidase (fibrin-stabilizing factor, coagulation factor XIIIa). This calcium-dependent enzyme was known to generate in fibrin clots covalent cross-links between adjacent fibrin molecules by coupling of specific glutamine side chains with suitably located ε-amino groups [39]. Fibronectin molecules could also be cross-linked by the same way [37].

If plasma fibronectin and fibrinogen were incubated with fibrin-stabilizing factor, thrombin, and 3.3 mM calcium chloride, complexes formed that, after reduction, exhibited a chain composition different from that of the starting materials as revealed by gel electrophoresis [37]. In addition to single peptide chains there were dimeric γ chains of fibrin, as well as high-molecular-weight products evidently representing covalent aggregates of fibronectin subunits and fibrin α chains. A semiquantitative determination showed a parallel decline of single fibrin α chains and fibronectin peptide chains during the reaction. In contrast, the γ chains of fibrin disappeared more rapidly with simultaneous production of γ-chain dimers, indicating a fast cross-linking reaction between these chains not influenced by fibronectin. Iwanaga and coworkers [31] demonstrated optimal covalent binding of [125 I] fibronectin to fibrin monomer-Sepharose by thrombin activated factor XIII in the presence of 5 mM calcium and 1 mM dithiothreitol.

Covalent cross-linking with fibrin can give rise to considerable binding of plasma fibronectin to a coagulum even at 37° C. In the absence of factor XIII or calcium ions, however, fibronectin remains in serum at 37° C (Table 2) [38].

Fibronectin, if covalently incorporated into fibrin fibrils, influences their elastic behavior [40]. The thin fibrils of a fine clot formed at high ionic strength and pH have half the elastic modulus, if cross-linked with fibronectin at 22° C, than fibronectin-free fibrils. In contrast, covalent binding of fibronectin in highly

Table 2 Plasma Fibronectin (%) in Serum Prepared by Coagulation of Plasma with Thrombin Under Various Conditions

Serum preparation		Fibronectin in serum (%)	References
Temperature (°C)	Calcium (mM)		
37	—	95-100	31, 38
37	40	45-53	38
0	—	35	31
2	40	<8	38

branched fibrils of a coarse clot formed at low pH and ionic strength increased the elastic modulus about twice compared with coarse fibrils lacking fibronectin.

The glutamine residues active in cross-linking can be labeled by transamidase-catalyzed coupling with suitable amines, such as putrescine, spermidine, or spermine [37]. Incorporation of [^{14}C] putrescine yielded a maximal uptake of 2.6 molecules per fibronectin, corresponding to somewhat more than 1 molecule per fibronectin subunit [41]. Using commercial plasma transamidase, Seidl [42] achieved a maximal incorporation if cysteine or other SH reagents were omitted, because these agents favored cross-linking of fibronectin in place of substitution.

The presence of more than one transamidase reactive site per fibronectin subunit also becomes evident, if incorporation of [^{14}C] putrescine is compared in fibronectin and in an N-terminal 70 kd-fragment containing the preferential cross-linking site (see below) [42, 43]. Although the uptake by the 70 kd-fragment was completed relatively soon, incorporation by fibronectin continued slowly over a longer period until a surplus of about 50% compared with the uptake of 70 kd-fragment was achieved (Fig. 1).

Location of Covalent and Noncovalent Fibrin Binding Sites in Fibronectin

Fibronectin consists of two high-molecular-weight subunits (molecular weight, about 250,000) connected by disulfide bonds close

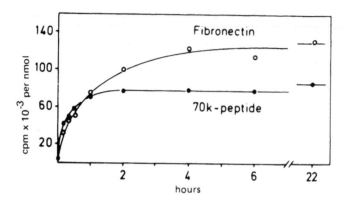

Figure 1 Incorporation of [^{14}C] putrescine into fibronectin and N-terminal
70 kd-peptide by crude plasma transamidase. Incorporation is recorded as
cpm per peptide chain. (From Ref. 42.)

to their C-terminal ends. Within the two subunits, distinct domains separable by proteolytic digestion can be discriminated [1-
4]. The composition of the two chains appears identical in most
of their parts. Only in the C-terminal region did some differences,
giving rise to a somewhat unequal length of the two chains, become evident [44-47].

The domain containing the main cross-linking site was primarily identified following degradation of fibronectin with plasmin
[48] (for other procedures, see Refs. 41 and 49). This enzyme
first cleaved the two subunit chains close to the interchain disulfide
bonds [50, 51]. Subsequently, a 29 kd-fragment was released, followed by a slow degradation of the residual chain. The 29 kd-fragment was identified as the N-terminal domain [52, 53]. Later investigations [45, 51] showed that a similar fragment was also released from the C-terminal part of the longer chain. The shorter
chain was devoid of the corresponding cleavage site.

Following plasminolysis of fibronectin, in which the transamidase reactive sites were substituted by [^{14}C] putrescine, most of
the incorporated radioactivity was found in the released 29 kd-
peptide [48]. The reactive site was correlated to a glutamine residue in position 3 subsequent to the N-terminal pyroglutamyl residue [54].

Evidently, this glutamine was the site responsible for covalent coupling of fibronectin and fibrin by transamidation, as electron micrographs of covalent fibrin-fibronectin complexes showed a connection of the two proteins at terminal positions [55]. On the other hand, one always found some [^{14}C] putrescine in the C-terminal part predominantly in the larger chain [44, 48]. Whether this minor transamidase reactive site has any meaning for cross-linking with fibrin is not known.

The N-terminal 29 kd-domain also contains a noncovalent affinity site for fibrin [56-58]. The domain is part of a cathepsin D-derived N-terminal 70 kd-fragment [59] with affinity to fibrin-Sepharose [56] and to gelatin-Sepharose [59]. Plasminolysis of the 70 kd-peptide yielded two fragments of 29 and 40 kd [59], the first exhibiting a fibrin binding site [56] and the second a gelatin binding site [59]. The fibrin-binding fragment was identified as the N-terminal piece as it contained radioactivity if prepared from [^{14}C] putrescine-labeled fibronectin [56]. The gelatin-binding 40 kd-peptide, evidently representing the subsequent domain was devoid of radioactivity.

When hamster fibronectin was degraded by thermolysin, Sekiguchi et al. [58] recognized a second fibrin binding site in addition to the terminal 29 kd domain. The second site was present in a 21 kd-fragment most likely released from a C-terminal region. The binding site was also found in a 200 kd-fragment generated by short trypsin degradation of fibronectin and was missing in a 180 kd fragment [57, 58]. The authors, therefore, concluded that only the larger subunit contained the second affinity site. Later experiments showed, however, that in human fibronectin both subunits expressed a fibrin-binding site in their C-terminal regions [46, 47]. It was readily released from the larger subunit chain by various proteases and appeared as a 37 kd-fragment. In contrast, the shorter chain resisted cleavage at that site and yielded, after prolonged digestion, a fibrin-binding 58 kd-fragment that also bound to heparin-Sepharose. Possibly, the 200 kd-fragment of hamster fibronectin also originated from the shorter subunit and the 180 kd-peptide represented an analogous piece of the longer chain appearing after release of a C-terminal fibrin binding domain.

Another location of the second fibrin binding site was proposed by Seidl and Hörmann [60]. Chymotrypsin digestion of fibronectin generated two fibrin binding fragments of 30 and 60 kd, the former identical with the N-terminal domain. The 60 kd-fragment was retained by gelatin-Sepharose as well and yielded the abovementioned gelatin-binding 40 kd-fragment and a nonbinding 18 kd-fragment if digested with cathepsin D. The 40 kd-fragment had only negligible affinity to fibrin and the 18 kd-fragment was inert, indicating nearly complete extinction of the binding site by the protease. As cathepsin D cleaves at the C-terminal end of the gelatin binding domain [53, 56], the fibrin binding site of the 60 kd peptide was correlated to its C-terminal region extending over the cathepsin D cleavage site. This fibrin affinity region, possibly overlapping with the gelatin binding domain, was in agreement with previous observations [36] that gelatin can prevent the absorption of fibronectin by a fibrin clot.

The results suggest that the fibronectin subunits might even contain three fibrin binding sites, as shown in Figure 2, although this is not yet certain. It is, however, interesting that all three possible regions exhibit a homologous structure. The N-terminal domain (fibrin binding site 1) contains five homologous sequences, each forming a disulfide stabilized loop (homology type I) [61]. Three loops of homology type I are also found in the C-terminal 37 kd domain released by plasmin (fibrin binding site 3) [62], as well as in the C-terminal part of the gelatin binding domain participating in the fibrin-binding site 2 [63]. In contrast, the amino terminal half of the gelatin binding domain contained two loops of homology type II [61, 62]. Although the N-terminal fibrin binding domain is positively charged, the two other domains with sequences of homology type I exhibit a negative net charge but are located in close proximity to positively charged regions [45, 60].

Fibronectin Binding Sites of Fibrin

Fibrinogen, the soluble precursor of fibrin, consists of three different peptide chains (Aα, Bβ, and γ) each present twice (for a review, see Ref. 64). As shown in Figure 3, the N-terminal regions of the six chains are connected to each other within the central part of the molecule by disulfide bonds and form there a globular domain (E domain). On both sides of this N-terminal disulfide knot protrude rodlike segments, each containing parts of three different peptide chains in a twisted α-helical conformation. Subsequently,

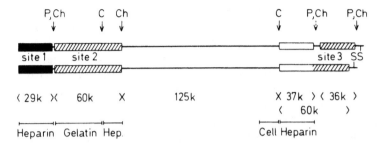

Figure 2 Location of fibrin binding sites within the fibronectin peptide chains. N-terminal site 1 (filled blocks) represents the main binding domain and contains the main transamidase reactive site. Proposed locations for a second and even a third affinity site (site 2 and 3) are labeled by hatched blocks. In the longer chain, site 3 is separated from an adjacent heparin-binding domain (open block) by a protease-sensitive peptide sequence that probably also bears a minor transamidase reactive site. In the shorter chain the two domains are close together. Early cleavage sites of cathepsin D (C), chymotrypsin (Ch), and plasmin (P), as well as affinity sites for heparin, gelatin, and cell surface receptors, are indicated.

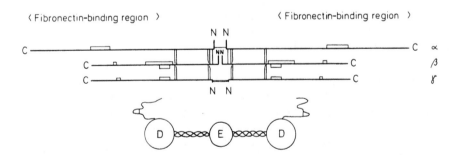

Figure 3 Fibrinogen. Scheme of peptide chain assembly with intrachain and interchain disulfide bonds (upper part). Trinodular structure with globular D and E domains connected by segments of coiled peptide chains (lower part). The flexible C-terminal sequences of the α chains are protruding from the two D domains.

on each side, outer globular domains (D domains) are located containing the C-terminal halves of the Bβ and γ chains in a disulfide-stabilized conformation. The C-terminal halves of the Aα chains exist in a more flexible conformation and show at most partial interaction with the other molecule (for a possible interaction with the E domain, see Ref. 65). Under the action of thrombin, short negatively charged fibrinopeptides A and B are released from the N-terminal regions of the Aα and Bβ chains, respectively. This altered charge distribution in the E domain induces an association of the formed fibrin monomers yielding fibrils.

Stathakis and Mosesson [23] had already shown that fibrinogen fractions of higher solubility (fraction I-8 and I-9 in the purification scheme of Mosesson and Sherry) [66], containing Aα chains more or less degraded on their C-terminal halves [67], were unable to form cryoprecipitable aggregates with fibronectin in the presence of heparin. Furthermore, mixtures of fibrin and fibrinogen prepared from those fractions by limited thrombin treatment failed to precipitate with fibronectin in the cold [34]. Evidently, the C-terminal parts of the fibrin α chains or fibrinogen Aα chains, respectively, are important for the affinity for fibronectin. It appears, however, that these regions of fibrin are more available for binding than the corresponding regions of fibrinogen, which is less effectively retained by fibronectin-Sepharose than fibrin monomer.

In analogous experiments plasma fibronectin was to a lesser degree adsorbed by fibrinogen-Sepharose than by fibrin-Sepharose prepared from the former by thrombin treatment (Fig. 4). On the other hand, immobilized α chains of fibrinogen and fibrin bound fibronectin equally well [42]. In these experiments, immobilized α chains were prepared by reductive cleavage of fibrinogen-Sepharose or fibrin-Sepharose, respectively. Earlier experiments had shown that, under these conditions, the β and γ chains were released and the α chains remained coupled to the Sepharose matrix [68]. The immobilized α chains retained the same chymotryptic fragments of fibronectin (29 and 60 kd) as fibrin-Sepharose [42].

The C-terminal halves of the α chains of fibrinogen and fibrin also interact with collagen, as shown by retention of fibrinogen by immobilized collagen, type III, in the cold [27]. Soluble fibrin was even retained by immobilized collagen, type I, at 4°C. In contrast, early plasminolysis products of fibrinogen had no affinity

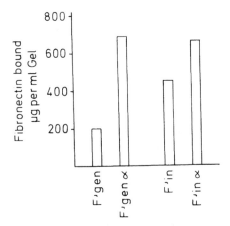

Figure 4 Retention of plasma fibronectin (5 mg) by columns (0.8 × 8 cm) filled with Sepharose-immobilized samples of fibrinogen (F'gen), fibrinogen Aα chains (F'gen α), fibrin monomer (F'in) and fibrin α chains (F'in α) at pH 7.6, ionic strength 0.15, 4°C. All absorbents had been prepared from the same batch of fibrinogen-Sepharose. (From Ref. 42.)

for either adsorbent. In this system, fibronectin might compete with collagen for the fibrin binding sites. Therefore, if in surgery fibrin adhesives are applied for gluing connective tissue, the preparations should be depleted of fibronectin, which probably weakens the fibrin-collagen interaction [69].

With regard to the covalent cross-linking region of fibrin for fibronectin, recently, various cyanogen bromide peptides derived from C-terminal parts of the α chains were obtained from polymerized fibrin in which fibronectin was incorporated [70, 71]. One of those fragments yielded several monoclonal antibodies, most of them directed against fibrin α chains. In addition, however, there was an antibody against a 30 kd fibronectin peptide indicating that this fragment, which presumably was the N-terminal domain, was cross-linked with fibrin [71].

INTERACTIONS ON THE CELL SURFACE

The affinity between fibronectin and fibrinogen-fibrin is essential for various processes on the cell surface. It concerns the attach-

ment of cells to a fibrin coagulum, the binding of soluble fibrin
by phagocytes as a requisite of internalization, or the shape changes
and aggregation of thrombocytes following activation. In most
cases the correlations are still incompletely understood and often
rely on indirect conclusions.

Macrophages and Cells of the Reticuloendothelial System

An important function of cells of the reticuloendothelial system is
the removal of soluble fibrin from circulating blood. Various findings provide evidence that fibronectin is involved in this process.
For instance, patients suffering from severe forms of disseminated
intravascular coagulation generally show a decreased level of plasma fibronectin [12].

Sherman and Lee [72] injected thrombin into rabbits that previously had received [^{125}I] fibronectin and [^{131}I] fibrinogen. As
a consequence the level of [^{125}I] fibronectin and [^{131}I] fibrinogen dropped, with similar kinetics and more rapidly than in untreated animals, which showed a [^{125}I] fibronection half-life of
about 72 hr. If ancrod was applied, a snake venom protease con-
verting fibrinogen into fibrin with degraded α chains, only defibrination without any decline of [^{125}I] fibronection was observed.
These results suggest that the decline of [^{125}I] fibronectin following thrombin infusion was due to a removal of fibrin-fibronectin
complexes.

Colvin and Dvorak [73] detected by immunofluorescence
methods fibrinogen-fibrin on the surface of peritoneal macrophages. They proposed fibronectin as a possible receptor because
both proteins showed a similar distribution on the cell surface.
Even capping experiments yielded parallel changes of the distribution pattern [74]. In experiments of Jilek and Hörmann [75], fibronectin improved the binding of [^{125}I] fibrin monomer to
macrophages. Gonda and Shainoff [76], on the other hand, investigated binding and internalization of fibrin electron microscopi-
cally and found evidence that the activated N terminus of fibrin
was recognized by the cells.

Fibroblasts

Grinnell et al. [10] reported in 1980 that attachment and spreading of fibroblasts on fibrinogen- or fibrin-coated surfaces required

fibronectin, factor XIII, and thrombin (for activation of factor XIII). In the absence of activated factor XIII, the cells attached less effectively and only a few cells spread. Apparently, the affinity between fibronectin and fibrin was insufficient at body temperature for mediating attachment and had to be supported by covalent cross-linking of the two proteins. Fibronectin with putrescine-blocked transamidase reactive sites failed to act as an attachment factor.

The results explain earlier observations of Niewiarowski et al. [11] that fibroblasts attach to polymerizing but not to cross-linked fibrin. Provided coagulation was initiated with thrombin, the fibroblasts induced a later contraction of the clot. Evidently, covalent cross-linking of polymerizing fibrin with fibronectin on the cell surface was essential for contraction. In this sytem a coagulum prepared with reptilase failed to contract, most likely because factor XIII was not activated. Virus-transformed fibroblasts with a reduced content of surface-bound fibronectin were ineffective in contracting a clot produced by thrombin [77].

Immunological methods revealed a codistribution of fibrinogen-fibrin and fibronectin on the surface of fibroblasts [78]. The binding of fibrinogen by the cells correlated with their fibronectin-binding capacity. The SV40-transformed and various malignant cells, which poorly bound fibronectin, failed to pick up fibrinogen. Otherwise, catabolic fibrinogen derivatives (fraction I-9) lacking the C-terminal halves of their α chains and, therefore, missing an affinity site for fibronectin, were bound by the cells far less effectively than normal fibrinogen. Although these experiments principally revealed no differences in the binding capacity of fibroblasts for fibrinogen and fibrin, it was mentioned that fibrin can polymerize on the cell surface and thus binding is improved.

Platelets

The role of fibronectin in platelet aggregation is the subject of Chapter 8, and only a short summary is given here.

If platelets are activated for aggregation, a pericellular structure develops on their surface involving participation of fibronectin [79], fibrinogen [80], and other compounds including von Willebrand's factor [81], thrombospondin [82], and probably heparin. All these substances are secreted by activated platelets,

114   HÖRMANN

although plasma components might also participate. Plow and
Ginsberg [83] demonstrated the generation of fibronectin receptors on the surface of activated platelets. Cell-bound fibronectin
and fibrinogen, bound by special fibrinogen receptors and/or associated with fibronectin, form bridges between cells, resulting in
their agglutination. In this process, thrombospondin, a lectinlike
protein, also comes into close contact with fibronectin [84].

CONCLUSIONS

Fibronectin has affinity to fibrinogen and fibrin, giving rise to
noncovalent associations. Complexes are well demonstrated in the
cold and dissociate more or less at body temperature. Each fibronectin subunit contains at least two, possibly three fibrin-binding
sites, which interact with the flexible C-terminal halves of the fi-
brin(ogen) α chains.

In addition to noncovalent interaction, fibronectin can covalently cross-link with fibrin through enzyme-catalyzed transamidation. In this reaction the C-terminal halves of the fibrin α
chains are coupled with a reactive glutamine residue close to the
N-terminus of the fibronectin subunit chains.

Interaction between fibronectin and fibrin-fibrinogen influences the coagulation process and the quality of the clot
formed. In addition, it is important for the attachment of adherent
cells on fibrin-containing structures, which is mediated by fibronectin. Covalent binding of cells to fibrin, probably by a cross-
link between cell-bound fibronectin and fibrin, appears to be a
prerequisite for contraction of a coagulum by included fibroblasts.
Finally, fibronectin might be important for removal of soluble fibrin from the circulating blood by cells of the reticuloendothelial
system, although further investigations are still required to settle
this point.

REFERENCES

1. D. F. Mosher, Fibronectin, in *Progress in Hemostasis and Thrombosis*
   (T. H. Spaet, ed.), Grune and Stratton, New York, 1980, pp. 111-151.
2. M. W. Mosesson and D. L. Amrani, The structure and biologic activities of
   plasma fibronectin, *Blood* 56:145-158 (1980).

3. E. Ruoslahti, E. Engvall, and E. G. Hayman, Fibronectin: Current concepts of its structure and functions. *Collagen Rel. Res.* 1:95-128 (1981).
4. H. Hörmann, Fibronectin—mediator between cells and connective tissue. *Klin. Wochenschr.* 60:1265-1277 (1982).
5. M. D. Pierschbacher, E. G. Hayman, and E. Ruoslahti, Location of the cell-attachment site in fibronection with monoclonal antibodies and proteolytic fragments of the molecule. *Cell* 26:259-267 (1981).
6. J. Molnar, S. McLain, C. Allen, H. Laga, A. Gara, and F. Gelder, The role of an α_2-macroglobulin of rat serum in the phagocytosis of colloidal particles. *Biochim. Biophys. Acta* 493:37-54 (1977).
7. F. A. Blumenstock, T. M. Saba, P. Weber, and R. Laffin, Biochemical and immunological characterization of human opsonic α_2-SB glycoprotein: Its identity with cold-insoluble globulin. *J. Biol. Chem.* 253:4287-4291 (1978).
8. J. E. Kaplan and P. W. Snedeker, Maintenance of fibrin solubility by plasma fibronectin. *J. Lab. Clin. Med.* 96:1054-1061 (1980).
9. J. Niewiarowska and C. S. Cierniewski, Inhibitory effect of fibronectin on the fibrin formation. *Thromb. Res.* 27:611-618 (1982).
10. F. Grinnell, M. Feld, and D. Minter, Fibroblast adhesion to fibrinogen and fibrin substrata: Requirement of cold-insoluble globulin (plasma fibronectin). *Cell* 19:517-525 (1980).
11. S. Niewiarowski, E. Regoezi, and J. F. Mustard, Adhesion of fibroblasts to polymerizing fibrin and retraction of fibrin induced by fibroblasts. *Proc. Soc. Exp. Biol. Med.* 140:199-204 (1972).
12. D. F. Mosher and E. M. Williams, Fibronectin concentration is decreased in plasma of severely ill patients with disseminated intravascular coagulation. *J. Lab. Clin. Med.* 91:729-735 (1978).
13. P. R. Morrison, J. T. Edsall, and S. G. Miller, Preparation and properties of serum and plasma proteins. XVIII. The separation of purified fibrinogen from fraction I of human plasma. *J. Am. Chem. Soc.* 70:3103-3108 (1948).
14. J. T. Edsall, G. A. Gilbert, and H. A. Scheraga, The nonclotting component of the human plasma fraction I-1 ("cold insoluble globulin") . *J. Am. Chem. Soc.* 77:157-161 (1955).
15. L. Thomas, R. T. Smith, and R. von Korff, Cold-precipitation by heparin of a protein in rabbit and human plasma. *Proc. Soc. Exp. Biol. Med.* 86:813-818 (1954).
16. R. T. Smith and R. W. von Korff, A heparin-precipitable fraction of human plasma. I. Isolation and characterization of the fraction. *J. Clin. Invest.* 36:596-604 (1957).
17. R. T. Smith, A heparin-precipitable fraction of human plasma. II. Occurrence and significance of the fraction in normal individuals and in various disease states. *J. Clin. Invest.* 36:605-616 (1957).

116 HÖRMANN

18. R. A. Heinrich, E. C. von der Heide, and A. R. W. Climie. Cryofrinogen: Formation and inhibition in heparinized plasma. *Am. J. Physiol. 204*: 419-422 (1963).
19. M. W. Mosesson and R. A. Umfleet, The cold-insoluble globulin of human plasma. I. Purification, primary characterization, and relationship to fibrinogen and other cold-insoluble fraction components. *J. Biol. Chem. 245:*5728-5736 (1970).
20. O. Fyrand, N. O. Solum, and P. Kierulf, Heparin precipitable fraction (HPF) from dermatological patients. I. Characterization of the thrombin-clottable protein. *Thromb. Res. 8:*17-29 (1976).
21. O. Fyrand and N. O. Solum, Heparin precipitable fraction (HPF) from dermatological patients. II. Studies on the non-clottable proteins. Identification of cold insoluble globulin as the main non-clottable component. *Thromb. Res. 8:*659-672 (1976).
22. M. Matsuda, T. Saida, and R. Hasegawa, Cryofibrinogen in the plasma of patients with skin ulcerative lesions on the legs. A complex of fibrinogen and cold-insoluble globulin. *Thromb. Res. 9:*541-552 (1976).
23. N. E. Stathakis and M. W. Mosesson, Interactions among heparin, cold-insoluble globulin, and fibrinogen in formation of the heparin-precipitable fraction of plasma. *J. Clin. Invest. 60:*855-865 (1977).
24. E. Engvall and E. Ruoslahti, Binding of soluble form of fibroblast surface protein, fibronectin, to collagen. *Int. J. Cancer 20:*1-5 (1977).
25. W. Dessau, F. Jilek, B. C. Adelmann, and H. Hörmann, Similarity of anti-gelatin factor and cold insoluble globulin. *Biochim. Biophys. Acta 533:* 227-237 (1978).
26. M. Vuento and A. Vaheri, Purification of fibronectin from human plasma by affinity chromatography under non-denaturing conditions. *Biochem. J. 183:*331-337 (1979).
27. F. Jilek and H. Hörmann, Affinity of fibrinogen and fibrin to collagen type III and denatured collagen type I demonstrated at low temperature. *Thromb. Res. 21:*265-272 (1981).
28. J. S. Finlayson and M. W. Mosesson, Heterogeneity of human fibrinogen. *Biochemistry 2:*42-46 (1963).
29. H. G. Klingemann, M. Kosukavak, H. Höfeler, and K. Havemann. Fibronectin and factor VIII related antigen in acute leukaemia. *Hoppe Seylers Z. Physiol. Chem. 364:*269-277 (1983).
30. E. Ruoslahti and A. Vaheri, Interaction of soluble fibroblast surface antigen with fibrinogen and fibrin. Identity with cold insoluble globulin of human plasma. *J. Exp. Med. 141:*497-501 (1975).
31. S. Iwanaga, K. Suzuki, and S. Hashimoto, Bovine plasma cold-insoluble globulin: Gross structure and function. *Ann. N. Y. Acad. Sci. 312:*56-73 (1978).

32. A. Stemberger and H. Hörmann, Heterogeneity of clottable fibrinogen isolated from plasma by affinity chromatography. *Hoppe Seylers Z. Physiol. Chem. 356*:341-348 (1975).

33. A. Stemberger and H. Hörmann, Affinity chromatography on immobilized fibrinogen and fibrin monomer. II. The behavior of cold-insoluble globulin. *Hoppe Seylers Z. Physiol. Chem. 357*:1003-1005 (1976).

34. N. E. Stathakis, M. W. Mosesson, A. B. Chen, and D. K. Galanakis, Cryoprecipitation of fibrin-fibrinogen complexes induced by cold-insoluble globulin of plasma. *Blood 51*:1211-1222 (1978).

35. D. F. Mosher and R. B. Johnson, Specificity of fibronectin-fibrin crosslinking. *Ann. N. Y. Acad. Sci. 408*:583-593 (1983).

36. E. Engvall, E. Ruoslahti, and E. J. Miller, Affinity of fibronectin to collagens of different genetic types and to fibrinogen. *J. Exp. Med. 147*: 1584-1595 (1978).

37. D. F. Mosher, Cross-linking of cold-insoluble globulin by fibrin-stabilizing factor. *J. Biol. Chem. 250*:6614-6621 (1975).

38. D. F. Mosher, Action of fibrin-stabilizing factor on cold-insoluble globulin and α_2-macroglobulin in clotting plasma. *J. Biol. Chem. 251*: 1639-1645 (1976).

39. L. Lorand, Fibrinoligase: The fibrin-stabilizing factor system of blood plasma. *Ann. N. Y. Acad. Sci. 202*:6-30 (1972).

40. G. W. Kamykowski, D. F. Mosher, L. Lorand, and J. D. Ferry, Modification of shear modulus and creep compliance of fibrin clots by fibronectin. *Biophys. Chem. 13*:25-28 (1981).

41. D. F. Mosher, P. E. Schad, and J. M. Vann, Cross-linking of collagen and fibronectin by factor XIIIa. Localization of participating glutaminyl residues to a tryptic fragment of fibronectin. *J. Biol. Chem. 255*:1181-1188 (1980).

42. M. Seidl, Fibrinbindung und Domänenstruktur von Fibronektin. Doctor's Thesis, University of Munich, 1983.

43. H. Hörmann, F. Jilek, M. Seidl, and H. Richter, Location of transamidase sensitive sites and of fibrin binding regions within the domain structure of fibronectin, in *Factor XIII and Fibronectin* (R. Egbring and H. G. Klingemann, eds.). Medizinische Verlagsgesellschaft, Marburg, 1983, p. 175-180.

44. H. Richter, M. Seidl, and H. Hörmann, Location of heparin-binding sites of fibronectin. Detection of a hitherto unrecognized transamidase sensitive site. *Hoppe Seylers Z. Physiol. Chem. 362*:399-408 (1981).

45. H. Richter and H. Hörmann, Early and late cathepsin D-derived fragments of fibronectin containing the C-terminal interchain disulfide crosslink. *Hoppe-Seylers Z. Physiol. Chem. 363*:351-364 (1982).

46. M. Hayashi and K. M. Yamada, Domain structure of the carboxyl-terminal half of human plasma fibronectin. *J. Biol. Chem. 258*:3332-3340 (1983).

118 HÖRMANN

47. K. Sekiguchi and S. Hakomori, Domain structure of human plasma fibronectin. Differences and similarities between human and hamster fibronectins. *J. Biol. Chem.* 258:3967-3973 (1983).
48. F. Jilek and H. Hörmann, Cold-insoluble globulin. III. Cyanogen bromide and plasminolysis fragments containing a label introduced by transamidation. *Hoppe-Seylers Z. Physiol. Chem.* 358:1165-1168 (1977).
49. D. D. Wagner and R. O. Hynes, Topological arrangement of the major structural features of fibronectin. *J. Biol. Chem.* 255:4304-4312 (1980).
50. F. Jilek and H. Hörmann, Cold-insoluble globulin. II. Plasminolysis of cold-insoluble globulin. *Hoppe-Seylers Z. Physiol. Chem.* 358:133-136 (1977).
51. M. Kurkinen, T. Vartio, and A. Vaheri, Polypeptides of human plasma fibronectin are similar but not identical. *Biochim. Biophys. Acta* 624: 490-498 (1980).
52. M. B. Furie and D. B. Rifkin, Proteolytically derived fragments of human plasma fibronectin and their localization within the intact molecule. *J. Biol. Chem.* 255:3134-3140 (1980).
53. G. Balian, E. M. Click, and P. Bornstein, Location of a collagen-binding domain in fibronectin. *J. Biol. Chem.* 255:3234-3236 (1980).
54. R. P. McDonagh, J. McDonagh, T. E. Petersen, H. C. Thøgersen, K. Skorstengaard, L. Sottrup-Jensen, S. Magnusson, A. Dell, and H. R. Morris, Amino acid sequence of the factor XIIIa acceptor site in bovine plasma fibronectin. *FEBS Lett.* 127:174-178 (1981).
55. H. P. Erickson, N. Carrell, and J. McDonagh, Fibronectin molecule visualized in electron microscopy: A long, thin, flexible strand. *J. Cell Biol.* 9:673-678 (1981).
56. H. Hörmann and M. Seidl, Affinity chromatography on immobilized fibrin monomer. III. The fibrin affinity center of fibronectin. *Hoppe-Seylers Z. Physiol. Chem.* 361:1449-1452 (1980).
57. K. Sekiguchi and S. Hakomori, Identification of two fibrin-binding domains in plasma fibronectin and unequal distribution of these domains in two different subunits. *Biochem. Biophys. Res. Commun.* 97:709-715 (1980).
58. K. Sekiguchi, M. Fukuda, and S. Hakomori, Domain structure of hamster plasma fibronectin. Isolation and characterization of four functionally distinct domains and their unequal distribution between two subunit polypeptides. *J. Biol. Chem.* 256:6452-6462 (1981).
59. G. Balian, E. M. Click, E. Crouch, J. M. Davidson, and P. Bornstein, Isolation of a collagen-binding fragment from fibronectin and cold-insoluble globulin. *J. Biol. Chem.* 254:1429-1432 (1979).
60. M. Seidl and H. Hörmann, Affinity chromatography on immobilized fibrin monomer. IV. Two fibrin-binding peptides of a chymotryptic digest of human plasma fibronectin. *Hoppe-Seylers Z. Physiol. Chem.* 364:83-92 (1983).

61. K. Skorstengaard, H. C. Thøgersen, K. Vibe-Pedersen, T. E. Petersen, and S. Magnusson, Purification of twelve cyanogen bromide fragments from bovine plasma fibronectin and the amino acid sequence of eight of them. Overlap evidence aligning two plasmic fragments, internal homology in gelatin-binding region and phosphorylation site near C-terminus. *Eur. J. Biochem.* *128*:605-623 (1982).

62. T. E. Petersen, H. C. Thøgersen, K. Skorstengaard, K. Vibe-Pedersen, P. Sahl, L. Sottrup-Jensen, and S. Magnusson, Partial primary structure of bovine plasma fibronectin: Three types of internal homology. *Proc. Natl. Acad. Sci. USA* *80*:137-141 (1983).

63. K. Skorstengaard, personal communication, 1982.

64. A. Henschen, F. Lottspeich, M. Kehl, and C. Southan, Covalent structure of fibrinogen. *Ann. N. Y. Acad. Sci.* *408*:28-43 (1983).

65. M. W. Mosesson, J. Hainfeld, J. Wall, and R. H. Haschemeyer, Identification and mass analysis of human fibrinogen molecules and their domains by scanning transmission electron microscopy. *J. Mol. Biol.* *153*:695-718 (1981).

66. M. W. Mosesson and S. Sherry, The preparation and properties of human fibrinogen of relatively high solubility. *Biochemistry* *5*:2829-2835 (1966).

67. M. W. Mosesson, J. S. Finlayson, R. A. Umfleet, and D. Galanakis, Human fibrinogen heterogeneities. I. Structural and related studies of plasma fibrinogens which are high solubility catabolic intermediates. *J. Biol. Chem.* *247*:5210-5219 (1972).

68. F. R. Matthias, D. C. Heene, and Z. Wegrzynowicz, Reduction of insolubilized fibrinogen. *Thromb. Res.* *4*:803-808 (1974).

69. A. Stemberger, W. Hebeler, W. Duspiva, and G. Blümel, Fibrinogen cold-insoluble globulin mixtures as tissue adhesives. *Thromb. Res.* *12*:907-910 (1978).

70. J. H. Sobel, P. H. Ehrlich, S. Birken, A. J. Saffran, and R. E. Canfield, Monoclonal antibody to the region of fibronectin involved in cross-linking to human fibrin. *Biochemistry* *22*:4175-4183 (1983).

71. P. H. Ehrlich, J. H. Sobel, Z. A. Moustafa, and R. E. Canfield, Monoclonal antibodies to α-chain regions of human fibrinogen that participate in polymer formation. *Biochemistry* *22*:4184-4192 (1983).

72. L. A. Sherman and J. Lee, Fibronectin: Blood turnover in normal animals and during intravascular coagulation. *Blood* *60*:558-563 (1982).

73. R. B. Colvin and H. F. Dvorak, Fibrinogen/fibrin on the surface of macrophages: Detection, distribution, binding requirements, and possible role in macrophage adherence phenomena. *J. Exp. Med.* *142*:1377-1390 (1975).

74. R. B. Colvin, Fibrinogen-fibrin interactions with fibroblasts and macrophages. *Ann. N. Y. Acad. Sci.* *408*:621-633 (1983).

75. F. Jilek and H. Hörmann, Fibronectin (cold-insoluble globulin). V. Mediation of fibrin monomer binding to macrophages. *Hoppe-Seylers Z. Physiol. Chem. 359*:1603-1605 (1978).
76. S. R. Gonda and J. R. Shainoff, Adsorptive endocytosis of fibrin monomer by macrophages: Evidence of a receptor for the amino terminus of the fibrin α chain. *Proc. Natl. Acad. Sci. USA 79*:4565-4569 (1982).
77. B. Azzarone, C. Curatolo, G. Carloni, M. B. Donati, L. Morasca, and A. Macieira-Coelho, Fibrin clot retraction in normal and transformed avian fibroblasts. *INCI 67*:89-94 (1981).
78. R. B. Colvin, P. I. Gardner, R. O. Roblin, E. L. Verderber, J. M. Lanigan, and M. W. Mosesson, Cell surface fibrinogen-fibrin receptors on cultured human fibroblasts. Association with fibronectin (cold insoluble globulin, LETS protein) and loss in SV40 transformed cells. *Lab. Invest. 11*: 464-473 (1979).
79. R. O. Hynes, I. U. Ali, A. T. Destree, V. Mautner, M. E. Perkins, D. R. Senger, D. D. Wagner, and K. K. Smith, A large glycoprotein lost from the surfaces of transformed cells. *Ann. N. Y. Acad. Sci. 312*:317-342 (1978).
80. G. A. Marguerie, T. S. Edgington, and E. F. Plow, Interaction of fibrinogen with its platelet receptor as part of a multistep reaction in ADP-induced platelet aggregation. *J. Biol. Chem. 255*:154-161 (1980).
81. K. S. Sakariassen, P. A. Bolhuis, and J. J. Sixma, Human blood platelet adhesion to artery subendothelium is mediated by factor XIII–von Willebrand factor bound to the subendothelium. *Nature 279*:636-638 (1979).
82. T. K. Gartner, D. C. Williams, F. C. Minion, and D. R. Phillips, Thrombin-induced platelet aggregation is mediated by a platelet plasma membrane-bound lectin. *Science 200*:1281-1283 (1978).
83. E. F. Plow and M. H. Ginsberg, Specific and saturable binding of plasma fibronectin to thrombin stimulated human platelets. *J. Biol. Chem. 256*: 9477-9482 (1981).
84. J. Lahav, M. A. Schwartz, and R. O. Hynes, Analysis of platelet adhesion with a radioactive chemical cross-linking reagent: Interaction of thrombospondin with fibronectin and collagen. *Cell 31*:253-262 (1982).

7

Plasma Fibronectin and Fibrin Formation

Jan McDonagh, Masao Hada,* and Marek Kaminski
Beth Israel Hospital
and Harvard Medical School
and the Charles A. Dana Research Institute
Boston, Massachusetts

Evidence is rapidly accumulating that there are multiple points of
interaction between the processes of coagulation, fibrinolysis, and
inflammation and that these are multiple facets of the general,
basic homeostatic mechanisms for repair and regeneration. One
point at which these processes may converge is through interaction
with fibronectin. There is substantial evidence that plasma fibronectin interacts with fibrin during polymerization and is incorporated into the fibrin gel [1]. Immunofluorescence studies have
shown that fibronectin is distributed along fibrin strands and raise

---

*Present affiliation: Department of Clinical Pathology, Tokyo Medical Col-
lege, Tokyo, Japan

the possibility of fibrin-fibronectin copolymerization [2]. The plasma concentration of fibronectin is about 320 µg/dl [3]. The serum concentration is reported to be 20–50% less than the plasma level and is due to incorporation of the protein into the clot [4]. Fibronectin is, in fact, the major protein in the clot after fibrin and constitutes 4–5% of the total protein [3]. An understanding of the effect of fibronectin on fibrin structure and function requires an understanding of fibrin formation. The mechanism for fibronectin interaction with fibrin must be within the framework of the basic mechanism of fibrin gel formation from fibrinogen.

CONVERSION OF SOLUBLE FIBRINOGEN TO A THREE-DIMENSIONAL GEL

Considerable information is available concerning the proteolytic activation of fibrinogen by thrombin and organization of the resultant fibrin monomers into protofibrils. Thrombin cleaves both A and B peptides, with the rate of cleavage of fibrinopeptide A being faster. Considerable evidence has been presented that these reactions are competitive [5, 6], although a sequential mechanism has also been proposed [7, 8]. In an interesting study on the cleavage rate of B peptide, Hurlet-Jensen et al. [9] used the tetrapeptide gly-pro-arg-pro to inhibit fibrin monomer aggregation and found that fibrinopeptide B was cleaved from fibrinogen and fibrin monomer at the same rate. Only when monomers were allowed to polymerize did the rate increase.

The substrate binding sites for these reactions have not been fully determined, although it is known that phenylalanine and aspartic acid in the A peptide (positions 8 and 7) are important [10]. This had led to the hypothesis that the conformation of the A peptide contains a hairpin like bend (type II β turn) stabilized by a salt bridge, which is inserted into the active-center pocket of thrombin [11]. Kinetic studies with fibrinogen fragments or synthetic peptides similar to the A peptide give values for k_{cat} that are reasonably close to intact fibrinogen, but they have poor values for K_m compared with fibrinogen [11]. Further studies with these fragments and conformationally specific antibodies toward the A and B peptide cleavage sites have indicated that only intact fibrinogen has the correct conformation for proper thrombin binding and that long-range interactions in fibrinogen are needed to maintain the proper substrate conformation [12, 13].

Once fibrinopeptide A cleavage has occurred, the resultant fibrin monomers spontaneously self-associate. It has been hypothesized that A peptide cleavage exposes a linear polymerization domain, and B peptide cleavage exposes a lateral association domain [14]. However, recent evidence indicates that the reactions are more complex. It is possible to achieve the same degree of lateral association without B peptide cleavage simply by slight alteration in the ionic strength [15]. The primary complementary domains involved in linear polymerization have been partly defined. One domain is located in the D region in the C-terminal part of the γ chain (Thr 374–Val 411); this domain is exposed in intact fibrinogen [16]. Its complementary binding site is located in the E region and involves the residues on the C-terminal side of the thrombin cleavage point of both the α and β chains. Synthetic peptides that mimic the new N-terminal regions of the α and β chains bind to the γ-chain peptide containing the polymerization domain [17, 18]. The β tetrapeptide but not the α shows enhanced binding in the presence of Ca^{2+} [19].

The first step in the organization of fibrin involves the assembly of fibrin monomers in a half-staggered overlap arrangement to form polymers two molecules thick, which are called protofibrils. This event involves the primary contact sites, just described, between the D and E domains, which have been referred to as DE-stag [20] and as A:a [21, 22]. Another set of contacts between adjacent D domains, called DD-long, also stabilizes the protofibrils [20]. The inhibition of gel formation by fibrin(ogen) degradation products occurs at the protofibril formation stage [23]. Lateral association between protofibrils serves to increase the fiber diameter, and the rate of this reaction determines the final fiber diameter [24, 25]. At the present time little is known about the mechanism of lateral association or the subsequent network branching of the fibers. Nevertheless, this model of fibrin assembly leads to three conclusions supported by considerable experimental evidence and important for understanding the interactions of fibrin molecules with other molecules. (1) Fibrin assembly is an orderly process initiated only through the addition of monomers to monomers or to the ends of growing protofibrils. (2) Lateral association occurs between protofibrils, not between monomers and protofibrils. (3) Fibrin monomers, and therefore protofibrils, are "sided" with respect to binding domains, so that the protofibril is closed with respect to linear polymerization sites along the polymer and only lateral sites are exposed.

During the process of assembly of the fibrin gel, factor XIII is activated by thrombin and Ca^{2+} and begins to reinforce the gel network through covalent cross-linking. Factor XIIIa (plasma transglutaminase) forms ϵ-(γ-glutamyl)lysyl cross-links between γ chains and α chains of fibrin. A reactive glutamine and lysine is in each γ chain (Glu-397 and Lys-405); these are cross-linked to another γ chain in an antiparallel manner to form a γ dimer [26]. γ Dimer formation occurs very rapidly and is probably complete before fibrin assembly has advanced beyond the protofibril stage. The cross-linking runs linearly through the γ chains along the protofibril strand and does not occur between the two strands of the protofibril [27]. Cross-linking of α chains is slower and builds a complex polymer. The reactive α-chain glutamines are at positions 326 and 366, and the lysines are between residues 508 and 584 [28]. The geometry of α-polymer formation has not been determined. Polymers probably contain a minimum of seven α chains and may be of higher order.

Table 1 shows the incorporation of fibronectin into plasma clots prepared under various conditions. It can be seen that the concentration of Ca^{2+} in the plasma markedly affects the amount of fibronectin incorporated into the clot. This illustrates that fibronectin incorporation occurs primarily through covalent cross-linking to fibrin. The reaction is Ca^{2+} dependent because it is catalyzed by factor XIIIa. Fibronectin also exhibits some degree of noncovalent interaction with fibrin, which is enhanced at low temperature.

NONCOVALENT INTERACTION OF FIBRONECTIN WITH THE FIBRIN GEL

Fibronectin has two types of binding domains for fibrin(ogen) [29–31]. Noncovalent interaction with fibrin is stronger than with fibrinogen [29, 32]. Significant interaction with fibrinogen occurs only at 4°C, and fibronectin is also necessary for heparin precipita-

Table 1 Incorporation of Fibronectin into Fibrin Clots

Test system[a]	Maximum incorporation (%)
Plasma[b] + EDTA	10-12
Plasma	26
Plasma + $CaCl_2$	44-49
Fibrinogen-Fn[c] + EDTA	12
Fibrinogen-Fn + $CaCl_2$	90-92

[a]Each test sample was trace labeled with [^{125}I] fibronectin
and clotted with thrombin.
[b]Citrated plasma was used.
[c]Purified fibrinogen (2 mg/ml) and fibronectin (1 mg/ml) were
mixed in equal volumes. The fibrinogen contained sufficient
factor XIII for complete fibrin cross-linking.

tion of plasma fibrinogen [32, 33]. These precipitation reactions
require that fibrin and fibrinogen have intact α chains, indicating
that interaction occurs through the carboxyl-terminal half of the α
chain [32, 33]. Noncovalent interaction of fibronectin with fibrin
is less temperature dependent and can occur at room temperature,
although this reaction is also facilitated at 4°C. Intact fibronectin,
as well as various degradation fragments, have been reported to
bind to fibrin monomer-Sepharose both at room temperature and
at 4°C [3, 35, 36]. Fibronectin did not bind to fibrinogen-Sepharose at room temperature [35, 36].
    There is a fibrin binding domain in the amino-terminal region
of each chain of the dimeric fibronectin molecule. The amino-terminal segment of each chain is readily cleaved by several proteases
to yield a fragment of molecular weight 27,000-29,000, which
contains the principal fibrin binding domain [37] (see also Chap.
6). It will bind to fibrin monomer-Sepharose, although this interaction may also require interaction of the fragment with other fibronectin fragments [3]. There is also evidence for an additional
fibrin binding domain in the carboxy-terminal region of one of the
fibronectin subunits [38], possibly in both [30].
    The binding site on fibrinogen or fibrin monomer that parti-
cipates in noncovalent interaction with fibronectin has not been
well studied. Francis and Marder [39] have observed that fibro-

nectin coelutes with a small fibrinogen fraction that contains an elongated γ chain (M_r = 55,000) on ion-exchange chromatography. This fibrinogen fraction constitutes 3–6% of the total fibrinogen pool [39] and would therefore be sufficient to account for the noncovalent binding of fibronectin to fibrin. However, it has not been determined if this γ-chain variant contains a fibronectin binding site.

COVALENT INTERACTION BETWEEN FIBRONECTIN AND FIBRIN

The principal interaction of fibronectin with fibrin is covalent and occurs because both proteins are substrates for factor XIIIa (plasma transglutaminase). Factor XIIIa catalyzes the formation of covalent ϵ-(γ-glutamyl)lysyl bonds between γ chains of fibrin to form γ-γ dimers and between α chains of fibrin to form high-molecular-weight α polymers. Factor XIIIa also catalyzes the cross-linking of fibronectin to fibrin [1], fibronectin to collagen [40], and α_2-plasmin inhibitor to fibrin [41].

The primary cross-linking site in fibronectin was first located in the amino-terminal 27,000–30,000 plasmin cleavage fragment [37]. Factor XIIIa can cross-link lysine analogs, such as dansyl cadaverine and putrescine, to appropriate glutamine residues in substrate proteins, and this approach can be used to isolate the glutamines in a protein that participate in the cross-linking reaction. With this procedure, under oxidative conditions (no reduction of fibronectin disulfide bonds), purified, intact fibronectin covalently bound two molecules of lysine analog per molecule of fibronectin. On reduction one molecule was found in each fibronectin chain. Fibronectin, thus labeled, was digested with plasmin, and the labeled fragment was purified by affinity chromatography on gelatin-Sepharose. The label appeared in the amino-terminal 29,000 fragment. Further fragmentations with trypsin and thermolysin were carried out, and labeled fragments were purified to homogeneity and sequenced. The labeled amino acid was found to be the third residue from the amino terminals of the intact molecule with the sequence [Glu-Ala-*Gln*-Val-Gln-Pro-]. The position of the putrescine label attached to glutamine at position 3 was

confirmed by mass spectroscopy; there was no label on the glutamine at position 4 [42]. There is also an additional, minor cross-
linking glutamine site, probably in the same region as the additional fibrin binding domain [30, 38].

Cross-linking of fibronectin to fibrin occurs through the α chain
of fibrin. This was shown by isolation of the reduced, alkylated fibronectin-fibrin complex, followed by partial sequence analysis.
Only α chain and fibronectin were found in the covalent complex
[36]. When fibrin and fibronectin samples in which lysine residues
were amidinated were used, it was found that the lysine residues in
the cross-linking reaction were contributed by fibrin and the glutamines were in fibronectin [3]. There are five lysine residues in
the α chain, located between residues 518 and 584, that can participate in covalent cross-linking [28]. It is not known which of
these preferentially cross-links to fibronectin. However, it is clear
that fibronectin is attached to the carboxy-terminal tail of the α
chain through its amino-terminal end. It is possible to prepare in
vitro fibronectin-fibrin clots that are soluble in dispersing solvents
if the fibronectin concentration in the clotting mixture is sufficiently high, indicating that fibrin cross-linking has been inhibited
[43]. In our experience this requires a 20-fold higher fibronectin-
fibrinogen ratio than occurs normally in plasma. When the molar
ratio of fibronectin to fibrinogen was 1.9:1.0, complete solubility
of the clot was observed in monochloroacetic acid. At 0.96:1.0,
partial solubility was seen, and at 0.78:1.0 the clot was insoluble,
indicating a sufficient degree of fibrin cross-linking to stabilize the
gel. These results indicate that high-molecular-weight copolymers
of α chain and fibronectin do not form.

STUDIES ON THE INCORPORATION OF FIBRONECTIN INTO FIBRIN GELS

Human fibronectin was purified from cryoprecipitate-depleted
Cohn fraction I or from human plasma by affinity chromatography on gelatin-Sepharose [44]. Trasylol was present in all
buffers throughout the purification procedure. The final product
was homogeneous on SDS gel electrophoresis and had a subunit
molecular weight of 220,000. It was radiolabeled with ^{125}I by the
solid phase lactoperoxidase-glucose oxidase method, using Enzyme-

beads (Bio-Rad). Labeled fibronectin was repurified on gelatin-
Sepharose to separate ligand from free iodine and also from radio-
lytically damaged material. Human fibrinogen was prepared from
the same starting materials by $(NH_4)_2SO_4$ precipitation [45]. It
was similarly labeled with ^{125}I or [131]I [46]. Fibronectin incorporation into fibrin clots was quantitated by electroimmunoassay
or by measuring radioactivity. Both methods gave the same results
($r^2 = 0.99$). In these studies radioactivity was measured in the fibronectin-fibrinogen solutions and in the separated clots and super-
natants. Total recovery of radioactivity after clotting was greater
than 95%. In some experiments fibronectin incorporation was
assessed by autoradiography or by slicing gels into 1 mm lateral
sections after electrophoresis and counting the sections. Factor
XIII, used for cross-linking fibronectin to fibrin, was purified from
plasma or placental concentrate (Behringwerke), as previously described [47], and was assayed by incorporation of [3H]putrescine
into casein [48].

In Figure 1 the incorporation of fibronectin into fibrin clots
was monitored by electroimmunoassay to compare the fibronectin
concentration of the solution before clotting to that in the supernatant 30 min after clotting had occurred. The linear range for
measurement of fibronectin concentration by this method was
found to be 25–500 μg/ml. In this experiment the fibrinogen concentration was constant at 400 μg/ml and the fibronectin concentration varied from 20 to 200 μg/ml. Clotting was initiated by
thrombin in the presence of sufficient $CaCl_2$ and factor XIII to
permit complete fibrin cross-linking. It can be seen that, in all samples, fibronectin was largely incorporated into the clots, ranging
from 97% incorporation at the lowest mole ratio of fibronectin to
fibrinogen (1:25) to 81% at 1:2.5 fibronectin-fibrinogen. The time
course of fibrinogen incorporation was also monitored. When the
fibrinogen concentration was 0.8 mg/ml and fibronectin was 0.4
mg/ml, incorporation was essentially complete 3 min after clotting
was initiated with 1 U/ml thrombin. At 45 sec, 79% of the fibronectin was incorporated into the clot, at 3 min, 88%, and at 3 min,
89%. All these studies indicated that conditions could be established in purified systems containing fibrinogen, fibronectin, factor XIII, thrombin, and $CaCl_2$ under which fibronectin was readily
incorporated into fibrin clots. Under optimal conditions approxi-

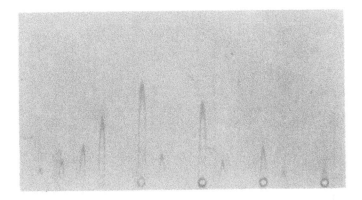

Figure 1 Quantitation of fibronectin before and after clotting by electroimmunoassay. The first four wells contain constant fibrinogen (400 μg/ml) and increasing fibronectin (12.5, 25, 50, and 100 μg/ml). Wells 5-12 are paired samples assayed for fibronectin before clotting (odd numbers) and in the supernatant after clotting (even numbers). The fibronectin concentrations before clotting were 200 μg/ml in 5 and 6, 140 μg/ml in 7 and 8, 60 μg/ml in 9 and 10, and 20 μg/ml in 11 and 12.

mately 90% of the fibronectin became incorporated into the clot over a wide range of fibronectin-fibrinogen mole ratios, up to a ratio of approximately one fibronectin molecule per two fibrinogen molecules.

Figure 2 shows the time course of fibronectin incorporation and cross-linking to fibrin in a purified system. At various times after clotting was initiated, clots were removed, washed in EDTA buffer, and dissolved in urea-sodium docecyl sulfate-dithiothreitol for gel electrophoresis. After electrophoresis in 4% polyacrylamide, the gels were sliced in 1 mm sections, which were then counted in a γ counter. At the early time points fibronectin appeared to be cross-linked to fibrin α chain to form heterodimers containing one fibronectin polypeptide subunit and one α chain. Significant amounts of higher molecular weight oligomers were observed up to 30 min, by which time most of the fibronectin was in the high-molecular-weight α-polymer material.

In contrast to the significant level of fibronectin incorporation into fibrin clots in purified systems, even at high fibronectin concentrations, markedly less fibronectin is incorporated into clots

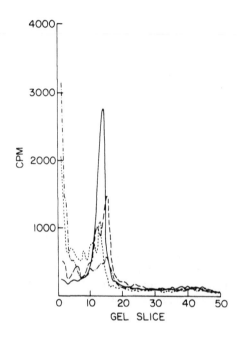

Figure 2 Time course of fibronectin cross-linking to fibrin. Samples containing fibrinogen, [125I] fibronectin, and factor XIII were clotted with thrombin and $CaCl_2$. At various times the reaction was stopped by immersing the
clot in urea-SDS-DTT-buffer and heating. Samples were electrophoresed in
SDS gel under reducing conditions; gels were then sliced in 1 mm sections and
counted in a γ-counter. (——): clotting stopped after 45 sec; (— —): 3 min;
(- - - -): 15 min; (—·—·): 30 min.

prepared from plasma. Figure 3 shows the time course of fibronectin incorporation in a plasma clot when clotting was initiated by
addition of thrombin and $CaCl_2$ to plasma containing trace-labeled
[125I] fibronectin. Even though the fibronectin-fibrinogen mole
ratio was approximately 1:10, only 46% of the total fibronectin
could be incorporated into the clot even after 2 hr incubation. We
have repeatedly observed that, with various in vitro test conditions
in plasma, the maximum amount of fibronectin that can be incorporated into fibrin clots is in the range of 40-50% of the total. This
is also in agreement with observations of other investigators [4].

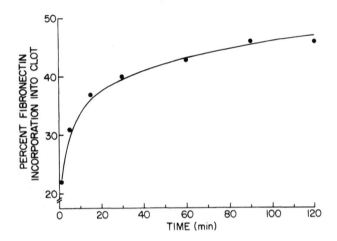

Figure 3 Time course of fibronectin cross-linking in a plasma clot. Plasma to which [^{125}I] fibronectin was added was clotted with thrombin and CaCl$_2$. At various times EDTA was added to stop the reaction, and the clot was removed, washed, and counted.

The rate of incorporation in plasma is also slower than in purified systems.

The reason for this discrepancy is not immediately obvious, and several reasonable possibilities exist, including competing proteolysis of fibronectin during plasma clotting, competition between fibronectin and other plasma proteins for fibrin binding and crosslinking, or a rate-limiting step in plasma that has been obviated in the purified system. Figure 4 shows autoradiographic patterns for fibronectin during plasma clotting. [^{125}I] fibronectin was added to plasma, which was then clotted with thrombin and CaCl$_2$ for various times. Clots were removed, washed, and solubilized. Aliquots of dissolved clots and serum samples were electrophoresed on agarose-acrylamide gels, and autoradiographs were prepared. The autoradiographs clearly demonstrate two important points. The first is that, not only is incorporation of fibronectin slower in plasma, but the rate of formation of high-molecular-weight polymers is significantly reduced. Cross-linked oligomers containing fibronectin were present in the plasma clots, as indicated by faint bands and

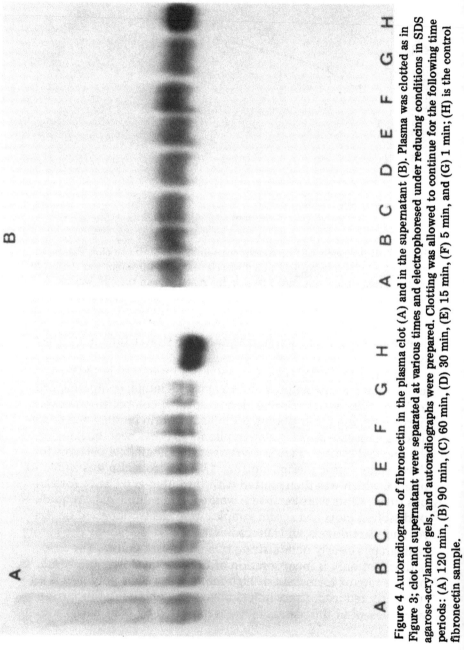

Figure 4 Autoradiograms of fibronectin in the plasma clot (A) and in the supernatant (B). Plasma was clotted as in Figure 3; clot and supernatant were separated at various times and electrophoresed under reducing conditions in SDS agarose-acrylamide gels, and autoradiographs were prepared. Clotting was allowed to continue for the following time periods: (A) 120 min, (B) 90 min, (C) 60 min, (D) 30 min, (E) 15 min, (F) 5 min, and (G) 1 min; (H) is the control fibronectin sample.

smudging in the upper regions of the gels, but even after 2 hr there was little distinctly visible high-molecular-weight polymer containing fibronectin. This is in contrast to the purified system in which essentially all the fibronectin was in the large polymer by 2 hr. The second important point is that the serum samples show no evidence that fibronectin was being degraded during the 2 hr clotting process. Thus, it appears unlikely that proteolysis could account for the failure of half the fibronectin to be incorporated into the plasma clot.

There is quite clearly a dependence on $CaCl_2$ for fibronectin incorporation, both in plasma and in purified systems (Table 1). However, this effect is maximal at 5 mM $CaCl_2$ in both systems. Since optimal concentrations were used in these studies, the Ca^{2+} requirement does not explain the differences in the fibronectin incorporation rates in the plasma and purified systems. Other plasma proteins that can be cross-linked to fibrin under certain circumstances include α_2-plasmin inhibitor [14] and von Willebrand's factor [49]. Tamaki and Aoki [50] have reported that fibronectin and α_2-plasmin inhibitor are cross-linked to the α chain of fibrin independently, and cross-linking of one does not influence the rate of cross-linking of the other. We have found similar results in comparing fibronectin and von Willebrand's protein cross-linking to fibrin; neither protein influences the interaction of the other [51]. Other plasma proteins that can interact with fibrin have not been studied in this way, and it is possible that, either singly or in combination, they might affect the rate of fibronectin incorporation. However, α_2-plasmin inhibitor is the third principal component of the plasma clot protein mass, after fibrin and fibronectin. Since it does not affect the rate of fibronectin cross-linking to fibrin, it seems unlikely that other lesser constituents would have a major effect.

Other factors that may influence the rate of fibronectin incorporation into the forming clot include temperature and ionic strength. Comparison of total fibronectin incorporation at room temperature and at $37°C$ showed that about 12% more fibronectin was cross-linked to the plasma clot at the higher temperature. With the purified system 8% more was incorporated at $37°C$ than at room temperature. When the ionic strength of the test system was increased from 0.15 to 0.30, approximately 9% more fibronectin was incorporated in the plasma clot at the higher ionic strength. With the purified system about 3% more was incorporated at higher ionic strength. Thus, with regard to effects of temperature and ionic strength on fibronectin incorporation, the two test systems are similar.

We have also investigated the effect of thrombin concentration on fibronectin incorporation into fibrin. These results are shown in Table 2. In all test samples $CaCl_2$ was added together with thrombin, and the clotting time was recorded. In one sample from each pair, the clot was removed at the time of gelation; in the other the clots were incubated for 60 min before being separated from the serum. All clots were washed, and the amount of fibronectin in them was determined. In can be seen that the amount of thrombin added had no effect on the incorporation of fibronectin into plasma clots. In fact, maximum incorporation occurred when no thrombin was added, and the plasma was clotted by recalcification and thrombin generation from plasma prothrombin. Approximately 50% of the maximum amount of fibronectin incorporation had occurred at the point of visible clot formation. In the purified system, in which the thrombin added constituted the total thrombin in the test system, fibronectin incorporation into the clot decreased with decreasing enzyme, indicating that the thrombin concentration had become rate limiting. When the clotting time was greater than 15 min, maximum fibronectin incorporation into the clot had decreased by 70%. However, in all samples 50% of the total fibronectin to be incorporated was in the clot at the time of gelation. These studies indicate that fibronectin incorporation into plasma fibrin and into purified fibrin gels occurs via the same mechanism.

Thrombin and factor XIIIa are the two enzymes responsible for fibrin formation and cross-linking and for cross-linking of fibronectin to fibrin. Factor XIII requires thrombin and Ca^{2+} for conversion to the active enzyme. Therefore, the effect of limiting thrombin on fibronectin incorporation in the purified system could reasonably be due either to an effect on the rate of protofibril formation or on the rate of activation of factor XIII. At the same time, the difference in the amount of fibronectin incorporated in the plasma clot could result from a rate-limiting concentration of factor XIIIa during plasma clotting. These possibilities

Table 2 Effect of Thrombin on the Incorporation of Fibronectin into
Fibrin Clots[a]

	Plasma test system			Purified test system		
Thrombin (U/ml)	Clotting time (sec)	% Incorporation A	B	Clotting time (min)	% Incorporation A	B
2.00	270	54	29	0.3	88	49
0.50	45	54	30	1.1	84	49
0.13	150	52	28	5.1	78	51
0.06	270	55	25	15.5	62	48

[a]In A the test system was incubated for 60 min after thrombin was added, and
then the clot and supernatant were separated. In B the clot was removed at the
time of clot formation.

were tested in several experiments. The test systems were modified
to use purified fibrinogen, treated with p-hydroxymercuribenzoate
(pCMB) to inactivate all factor XIII contamination in the preparation, and plasma from a patient congenitally deficient in factor
XIII with no measurable factor XIII activity. When a purified preparation of factor XIII, not containing fibrinogen or fibronectin,
was added to the test samples, the amount of fibronectin incorporated was found to be clearly a function of the factor XIII concentration (Fig. 5). This was observed in the purified system and in
normal and factor XIII-deficient plasma samples. When the factor
XIII concentration of normal plasma was increased twofold, there
was essentially no difference between the plasma system and the
purified system, indicating that the factor XIII concentration of
normal plasma is rate limiting for fibronectin cross-linking to fibrin in an in vitro clotting system. Similar results were obtained
with addition of factor XIII concentrate to XIII-deficient plasma.

The effect of factor XIII concentration on fibronectin incorporation was further confirmed by studying the time course of incorporation at several factor XIII concentrations (Fig. 6). It can be
seen that fibronectin incorporation increased as a function of factor XIII concentration with the expected time dependence. In an

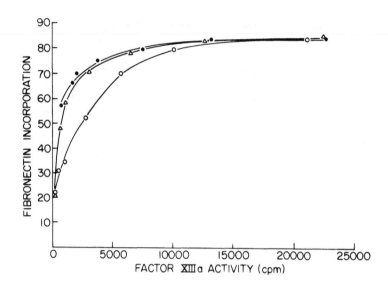

Figure 5 Relationship between Factor XIIIa concentration and fibronectin incorporation into the fibrin clot. Purified factor XIII was added to normal plasma (●), congenitally factor XIII-deficient plasma (△), and to purified fibrinogen plus fibronectin (○). Each sample was clotted with thrombin and $CaCl_2$, and the amount of fibronectin in the clot was measured. Factor XIIIa activity is expressed as the cross-linking of [^{14}C]putrescine to casein (cpm) measured in a separate assay.

additional experiment, fibronectin incorporation was measured when fibrinogen was clotted by the snake venom reptilase, which only releases fibrinopeptide A, not fibrinopeptide B. When pCMB-fibrinogen was clotted by reptilase in the presence of factor XIIIa, fibronectin incorporation was essentially normal (Fig. 7). For these experiments the activation peptide of factor XIII was cleaved by thrombin and thrombin was then inactivated by hirudin prior to admixture with fibrinogen. These experiments clearly indicate that cleavage of fibrinopeptide B has no effect on fibronectin cross-linking to the fibrin that subsequently forms. Therefore fibrin I, with fibrinopeptide A cleaved, is essentially as good a substrate for fibronectin cross-linking as fibrin II, with both fibrinopeptides removed.

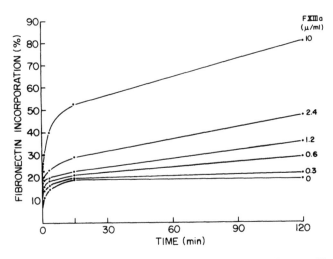

Figure 6 Time dependence of fibronectin cross-linking to fibrin as a function
of factor XIII concentration. Clotting mixture contained fibrinogen, [^{125}I]
fibronectin, and various amounts of factor XIII. Factor XIII concentration
was determined separately and is expressed as units per milliliter, where 1
U/ml is the activity of normal plasma.

All these experiments lead to the conclusion that the concentration of activated factor XIII is the principal factor regulating
the incorporation of fibronectin into fibrin gels. When fibrinogen
is activated through the proteolytic cleavage of fibrinopeptide A,
the resultant fibrin monomers will spontaneously aggregate to
form a three-dimensional gel. Whenever fibronectin is present in
the clotting solution, a certain small fraction, amounting to about
1 molecule of fibronectin per 100 fibrin molecules, will associate
with the fibrin gel through weak, noncovalent interaction. However, in the presence of active factor XIII, this interaction changes
dramatically. Factor XIIIa covalently cross-links fibronectin to the
α chain of fibrin; both the rate of this cross-linking reaction and
the total amount of fibronectin thus incorporated into the clot depend on the factor XIIIa concentration. The normal plasma concentration of factor XIII after activation is considered to be 1 U/ml.
In the clotting of normal plasma the molar cross-linking ratio of
fibronectin to fibrin is 1:20. If the factor XIIIa concentration of

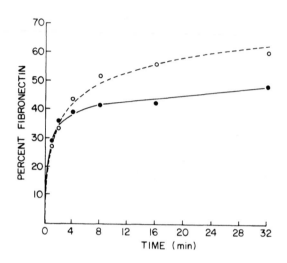

Figure 7 Incorporation of fibronectin into reptilase-fibrin clot. The test system contained factor XIII-free fibrinogen and fibronectin. Preactivated factor XIII (factor XIIIa) was added, and the mixture was clotted with reptilase and $CaCl_2$ (○) or thrombin and $CaCl_2$ (●).

plasma is increased to 2 U/ml, the mole ratio of fibronectin to fibrin rises to about 1:10, equivalent to essentially total cross-linking of fibronectin into the plasma clot. In a purified system with factor XIIIa at 10 U/ml, the ratio is 1 molecule of fibronectin per 2 molecules of fibrin.

The apparent dependence of fibronectin incorporation on the concentrations of Ca^{2+} and thrombin results from the requirement of these molecules for factor XIII activation and expression of activity. In an experimental system that does not contain factor XIII, increasing the Ca^{2+} or the thrombin concentration has no effect on fibronectin incorporation.

FIBRONECTIN INCORPORATION AND FIBRIN GEL STRUCTURE

Fibronectin cross-linking to fibrin is through the α chain of fibrin, but the step(s) in fibrin gel formation where fibronectin incorporation occurs is not entirely clear. Fibrin self-assembly is an ordered process, initiated by thrombin cleavage of fibrinopeptide A from fibrinogen. The resultant fibrin monomers aggregate in a half-staggered overlapping manner to form linear arrays two molecules in diameter, which are called protofibrils. Protofibrils associate laterally to form thicker fibrils, and through branching of the fibrils the three-dimensional gel is constructed. Factor XIIIa reinforces the gel structure by forming covalent bonds between pairs of γ chains and multimers of α chains. γ-Dimer formation occurs very rapidly, probably at the protofibril stage. α-Polymer cross-linking occurs much more slowly and is not complete until some time after visible clot formation. Our data (Table 2) indicate that approximately 50% of the fibronectin-fibrin cross-linking that will occur under several experimental conditions has already happened at the time of visible gel formation. The remaining 50% is then cross-linked more slowly to α-chain multimers. In other experiments we have also observed that fibronectin–α-chain cross-linking decreases the rate of α-chain cross-linking to form high-molecular-weight polymers [51].

Studies of fibronectin cross-linking to fibrin prepared from reptilase-treated fibrinogen show that fibrin containing fibrinopeptide B can be readily cross-linked to fibronectin (Fig. 7). Gels formed from fibrin monomers containing B peptide (fibrin I) differ from those with both peptides cleaved (fibrin II) principally in their elastic properties [15]. Assembly of both types of gels proceeds via the same mechanisms, but with fibrin I the contacts fit less well, and the gel is mechanically less strong than is fibrin II. It is therefore not surprising that both types of fibrin can cross-link to fibronectin. The thrombinlike enzymes, reptilase and ancrod, which cleave fibrinopeptide A but not B, also do not activate factor XIII. Hence, for fibronectin to be cross-linked to reptilase fibrin, it is necessary to have a source of preactivated factor XIII in the test system. This is probably the reason fibronectin interaction with fibrin was not observed in vivo in ancrod-treated animals, whereas it was in thrombin-treated animals [52].

The covalent cross-linking of fibronectin to fibrinogen to form heterodimers has not been carefully described but can probably occur, in a manner analogous to fibrinogen dimer formation [27], when the concentration of factor XIIIa is sufficiently high. Dimeric complexes of fibrinogen and fibronectin have been observed by electron microscopy [44] (see also Chap. 3), and the formation of circulating fibrinogen-fibrin-fibronectin complexes has been reported [53].

CONSEQUENCES OF FIBRONECTIN INTERACTION WITH FIBRIN

Fibronectin incorporation into the fibrin clot may affect several functional properties of the gel, including solubility, mechanical strength, and susceptibility to plasmin degradation. Some conflicting results have been reported, and at the present time it is difficult to understand the mechanism for all the reported effects. This may be due partly to the use of different experimental conditions by various investigators. The process of fibrin assembly is markedly influenced by pH, ionic strength, metal ion concentration, and enzyme and substrate concentrations, so that significant differences in gel structure (that is, fiber diameter, fiber length, and number of branch points) may arise, depending on the clotting conditions. It is possible that fibronectin might have variable effects on the fibrin gel depending on the conditions of polymerization. Both the molar ratios of fibronectin-fibrinogen and the absolute concentrations of the proteins may be important variables.

In test systems that were apparently lacking factor XIII activity, fibronectin has been reported to inhibit significantly fibrin monomer aggregation in dilute urea at a fibronectin-fibrin monomer mole ratio of 1:6 and also to dissolve noncovalently associated fibrin aggregates [54]. However, when the factor XIII concentration is sufficiently high, we have observed ready formation of a fibrin gel, even when the fibronectin-fibrinogen ratio is 1. Niewiarowska and Cierniewski [55] observed a concentration-dependent, inhibitory effect of fibronectin on the clotting time and on fibrin polymerization both in plasma and in a purified system. These studies were also apparently performed under conditions in which little or no covalent cross-linking occurred. In this study a significant effect of fibronectin on fibrin polymerization was observed at a fibronectin-fibrinogen ratio of 1:10. The lag phase increased and the maximum turbidity of the gel decreased with increasing fibronectin concentration. However, Galanakis and Simon [56], using similar test systems, have recently reported that fibronectin has no appreciable effect on fibrin aggregation.

When fibronectin is cross-linked to fibrin at molar ratios greater than 1, α-polymer formation is significantly inhibited and the gel can be dissolved in such solvents as urea or acetic acid. We interpret these results to mean that covalent, cross-linked, high-molecular-weight fibronectin polymers or fibronectin-fibrin co-polymers do not form. Such polymers would be theoretically possible since fibronectin is a dimeric, bifunctional molecule. Rather it appears that, in high concentration, fibronectin inhibits α-polymer formation by blocking the reactive lysine residues in the carboxy-terminal portion of the α chain.

The degree of α-polymer formation is an important determinant of the elasticity or mechanical strength of fibrin gels [57]. Since it causes a decrease in the degree of α-polymer formation, fibronectin-fibrin cross-linking should concomitantly lower the elasticity. In addition, fibronectin incorporation may have other, more complex effects on gel elasticity. Using a test system in which the mole ratio of fibronectin to fibrinogen was approximately 1:7, Kamykowski et al. [58] found that, at physiological ionic strength and pH, fibronectin incorporation results in a 2.5-fold increase in gel elasticity. In these experiments about one-third of the α chains formed cross-linked α polymer in the absence of fibronectin. However, at high ionic strength and pH, conditions that favor the formation of thin fibers, fibronectin incorporation lowered the gel elasticity twofold. These experiments were done under conditions that resulted in complete γ-dimer formation and no α-polymer formation in the control sample without fibronectin [58]. The markedly different effects of fibronectin in thin fiber versus thick fiber gels are somewhat difficult to understand but may relate to differences in polymerization kinetics in the two systems. Chou et al. [59] measured the mechanical properties of fibrin clots formed in recalcified plasma and found that the elasticity decreased by 3% per 0.1 mg/ml fibronectin incorporation. These experiments were carried out under conditions in which factor XIII activity should be minimal, although this was not quantitated. The effect of fibronectin incorporation on gel elasticity has apparently not been determined at optimal concentrations of factor XIIIa or with optimal conditions for α-polymer formation.

The potential effects of fibronectin on fibrinolysis may also be complex. Covalent cross-linking reactions are important in regulating the susceptibility of the fibrin gel to plasmin degradation. This occurs through the cross-linking of fibrin, principally in α-polymer formation [60], through the cross-linking of α_2-plasmin inhibitor to fibrin α chains, and also possibly through differential binding of plasminogen to cross-linked and non-cross-linked fibrins [41]. Any factor that alters these cross-linking reactions could change the susceptibility of the fibrin gel to degradation. Tamaki and Aoki [50] have shown that α_2-plasmin inhibitor and fibronectin are cross-linked to fibrin α chains independently, and one does not influence the cross-linking rate of the other [50]. However, by decreasing the rate of α-polymer formation, fibronectin might increase the susceptibility of the gel to lysis. Such results were obtained by Iwanaga et al. when urokinase was used to activate plasminogen in the clot. Increasing concentrations of fibronectin resulted in shorter clot lysis times. In that system a gel formed from a 1:30 mixture of fibronectin and fibrinogen lysed in about 55 min, compared with a lysis time of 100 min in the control gel without fibronectin [36]. In preliminary experiments in which gels were formed from equimolar amounts of fibrinogen and fibronectin, we have observed that the initial rates of fibrin degradation were unaffected by fibronectin. Fibronectin in the clot was degraded more rapidly than fibrin, as monitored by the appearance of radiolabeled fragments in the supernatant. Iwanaga et al. [36] have also reported that fibronectin or fibronectin fragments enhance the rate of plasminogen activation by urokinase.

From the foregoing discussion it seems reasonable that fibronectin may have several complex effects on fibrin gel structure and function. In addition to direct alterations of gel structure that may change clot solubility, mechanical stability, and sensitivity to lysis, fibronectin also changes the substrate specificity of fibrin for adherence of fibroblasts. Fibrin containing cross-linked fibronectin is a far better substratum for cell attachment and spreading than is fibrin alone [61]. Fibronectin mediates the opsonization of denatured collagen by macrophages and may also be important in fibrin attachment to macrophages [62] (see also Chaps. 9 and 10). Both of these fibrin cell-associated functions may be important for normal wound healing. Fibronectin can also be covalently cross-linked to collagen. Whether trimolecular collagen-fibronectin-fibrin cross-linked complexes can be formed has not been determined. Such a mechanism could be effective in vivo in anchoring the gel at the wound site and in increasing its resistance to blood flow.

PLASMA FIBRONECTIN IN VIVO

The average fibronectin concentration of normal plasma is about 300–320 μg/ml, with a range of 180–720 μg/ml [63]. There is some age and sex dependence [64]. In studies in vitro about 50% of the fibronectin is cross-linked to fibrin when plasma is clotted. It is, however, not known how much fibronectin is cross-linked to fibrin during in vivo clot formation. In vivo there are other potential sources of factor XIII activity in platelets and monocytes that could increase the local concentration at the site of clot formation [65, 66]. In addition, there are tissue transglutaminases, which can cross-link fibronectin [67]. These could also be active in a wound site. Thus, it seems likely that there is sufficient factor XIII activity available for fibronectin to have a modulatory role in fibrin formation and degradation in vivo.

Plasma fibronectin levels are generally low to very low in patients with overt disseminated intravascular coagulation (DIC) [63]. In other severely ill patients, for example with advanced cancer or sepsis, the levels vary throughout the normal range and no clear correlation has been established [64]. Plasma fibronectin is synthesized in the liver and secreted into the circulation [68]. Patients with obstructive liver disease generally have higher than normal levels [64]. The consequences of decreased fibronectin levels are not yet clear, although clinical improvement has been observed after infusion of cryoprecipitate in severely ill patients with DIC [69, 70]. Although in vitro studies indicate an important function for fibronectin in regulating fibrin formation and degradation, more studies, both in patients and in experimental systems, are needed to determine the mechanisms of in vivo function and the possible consequences of abnormally high or low levels in the pathogenesis of various disease processes.

144   MCDONAGH, HADA, AND KAMINSKI

REFERENCES

1. D. F. Mosher, Cross-linking of cold-insoluble globulin by fibrin-stabilizing factor. *J. Biol. Chem. 250*:6614-6621 (1975).
2. F. Grinnell and M. K. Feld, Distribution of fibronectin on peripheral blood cells in freshly clotted blood. *Thromb. Res. 24*:397-404 (9181).
3. D. F. Mosher and R. B. Johnson, Specificity of fibronectin-fibrin crosslinking. *Ann. N. Y. Acad. Sci. 408*:583-593 (1983).
4. M. W. Mosesson and R. A. Umfleet, The cold-insoluble globulin of human plasma. *J. Biol. Chem. 245*:5728-5736 (1970).
5. R. A. Martinelli and H. A. Scheraga, Steady-state kinetics study of the bovine thrombin-fibrinogen interaction. *Biochemistry 19*:2343-2350 (1980).
6. H. L. Nossel, A. Hurlet-Jensen, C. Y. Liu, J. A. Koehn, and R. E. Canfield, Fibrinopeptide release from fibrinogen. *Ann. N. Y. Acad. Sci. 408*: 269-277 (1983).
7. B. Blombäck, A. Hessel, D. Hogg, and L. Therkildsen, A two-step fibrinogen-fibrin transition in blood coagulation. *Nature 275*:301-304 (1978).
8. S. D. Lewis, D. L. Higgins, and J. A. Shafer, Effects of EDTA, Gly-pro-arg-pro, and amino acid replacement on the thrombin-catalyzed release of fibrinopeptides. *Ann. N. Y. Acad. Sci. 408*:669-670 (1983).
9. A. Hurlet-Jensen, H. Z. Cummins, H. L. Nossel, and C. Y. Liu, Fibrin polymerization and release of fibrinopeptide B by thrombin. *Thromb. Res. 27*:419-427 (1982).
10. H. C. March, Jr., Y. C. Meinwald, S. Lee, and H. A. Scheraga, Mechanism of action of thrombin on fibrinogen. Direct evidence for the involvement of phenylalanine at position P$_9$. *Biochemistry 21*:6167-6171 (1982).
11. H. A. Scheraga, Interaction of thrombin and fibrinogen and the polymerization of fibrin monomer. *Ann. N. Y. Acad. Sci. 408*:330-343 (1983).
12. J. A. Nagy, Y. C. Meinwald, and H. A. Scheraga, Immunochemical determination of conformational equilibria for fragments of the Aα chain of fibrinogen. *Biochemistry 21*:1794-1806 (1982).
13. J. A. Nagy, Y. C. Meinwald, and H. A. Scheraga, Immunochemical determination of conformational equilibria for fragments of the Bβ chain of fibrinogen. *Biochemistry* submitted for publication.
14. T. C. Laurent and B. Blombäck, On the significance of the release of two different peptides from fibrinogen during clotting. *Acta Chem. Scand. 12*:1875-1877 (1958).
15. L. L. Shen, J. Hermans, J. McDonagh, and R. P. McDonagh, Role of fibrinopeptide B release: Comparison of fibrins produced by thrombin and ancrod. *Am. J. Physiol. 232*:629-633 (1977).
16. S. Olexa and A. Z. Budzynski, Localization of a fibrin polymerization site. *J. Biol. Chem. 256*:3544-3549 (1981).

17. A. P. Laudano and R. F. Doolittle, Synthetic peptide derivatives which bind to fibrinogen and prevent the polymerization of fibrin monomers. *Proc. Natl. Acad. Sci. USA* 75:3085-3089 (1978).
18. A. P. Laudano and R. F. Doolittle, Studies on synthetic peptides that bind to fibrinogen and prevent fibrin polymerization. Structural requirements, numbers of binding sites, and species differences. *Biochemistry* 19:1013-1019 (1980).
19. A. P. Laudano and R. F. Doolittle, Influence of calcium ion on the binding of fibrin amino terminal peptides to fibrinogen. *Science* 212:457-459 (1981).
20. W. E. Fowler, R. R. Hantgan, J. Hermans, and H. P. Erickson, Structure of the fibrin protofibril. *Proc. Natl. Acad. Sci. USA* 78:4872-4876 (1981).
21. B. J. Kudryk, D. Collen, K. R. Woods, and B. Blombäck, Evidence for localization of polymerization sites in fibrinogen. *J. Biol. Chem. 249*: 3322-3325 (1974).
22. B. Blombäck, B. Hessel, M. Okada, and N. Egberg, Mechanism of fibrin formation and its regulation. *Ann. N. Y. Acad. Sci. 370*:536-544 (1981).
23. J. E. Williams, R. R. Hantgan, J. Hermans, and J. McDonagh, Characterization of inhibition of fibrin assembly by fibrinogen fragment D. *Biochem. J. 197*:661-668 (1981).
24. R. R. Hantgan and J. Hermans, Assembly of fibrin: A light scattering study. *J. Biol. Chem. 254*:11272-11281 (1979).
25. R. R. Hantgan, J. McDonagh, and J. Hermans, Fibrin assembly. *Ann. N. Y. Acad. Sci. 408*:344-365 (1983).
26. R. Chen and R. F. Doolittle, γ-γ crosslinking sites in human and bovine fibrin. *Biochemistry 10*:4486-4491 (1971).
27. W. E. Fowler, H. P. Erickson, R. R. Hantgan, J. McDonagh, and J. Hermans, Electron microscopy of crosslinked fibrinogen dimers demonstrates a feature of the molecular packing in fibrin fibers. *Science 211*: 287-280 (1981).
28. B. A. Cottrell, D. D. Strong, K. W. K. Watt, and R. F. Doolittle, Amino acid sequence studies of the α chain of human fibrinogen. Exact location of crosslinking acceptor sites. *Biochemistry 18*:5405-5410 (1979).
29. H. Hörmann and M. Seidl, Affinity chromatography on immobilized fibrinogen and fibrin monomer. *Hoppe-Seylers Z. Physiol. Chem. 361*: 1449-1452 (1980).
30. H. Richter, M. Seidl, and H. Hörmann, Location of heparin-binding sites of fibronectin. Detection of a hitherto unrecognized transamidase sensitive site. *Hoppe-Seylers Z. Physiol. Chem. 362*:399-408 (1981).
31. K. Sekiguchi and S. Hakamori, Domain structure of human plasma fibronectin. Differences and similarities between human and hamster fibronectins. *J. Biol. Chem. 258*:3967-3973 (1983).

32. N. E. Stathakis, M. W. Mosesson, A. B. Chen, and D. K. Galanakis, Cryo-precipitation of fibrin-fibrinogen complexes induced by the cold-insoluble globulin of plasma. *Blood 51*:1211–1222 (1978).
33. N. E. Stathakis and M. W. Mosesson, Interactions among heparin, cold-insoluble globulin, and fibrinogen in formation of the heparin-precipitable fraction of plasma. *J. Clin. Invest. 60*:855–866 (1977).
34. E. Ruoslahti and A. Vaheri, Interaction of soluble fibroblast surface antigen with fibrinogen and fibrin. Identity with cold insoluble globulin of human plasma. *J. Exp. Med. 141*:497–501 (1975).
35. A. Stemberger and H. Hörmann, Affinity chromatography on immobilized fibrinogen and fibrin monomer. II. The behavior of cold-insoluble globulin. *Hoppe-Seylers Z. Physiol. Chem. 357*:1003–1005 (1976).
36. S. Iwanaga, K. Suzuki, and S. Hashimoto, Bovine plasma cold-insoluble globulin: Gross structure and function. *Ann. N. Y. Acad. Sci. 312*:56–73 (1978).
37. F. Jilek and H. Hörmann, Cold-insoluble globulin. III. Cyanogen bromide and plasminolysis fragments containing a label introduced by transamidation. *Hoppe-Seylers Z. Physiol. Chem. 358*:1165–1168 (1977).
38. K. Sekiguchi and S. Hakamori, Identification of two fibrin binding domains in plasma fibronectin and unequal distribution of these two domains in two different subunits. *Biochem. Biophys. Res. Commun. 97*: 709–715 (1980).
39. C. W. Francis and V. J. Marder, Heterogeneity of normal human fibrinogen due to two high molecular weight variant γ chains. *Ann. N. Y. Acad. Sci. 408*:118–120 (1983).
40. D. F. Mosher, P. E. Schad, and H. K. Kleinman, Cross-linking of fibronectin to collagen by blood coagulation factor XIIIa. *J. Clin. Invest. 64*: 781–787 (1979).
41. Y. Sakata and N. Aoki, Crosslinking of α_2-plasmin inhibitor to fibrin by fibrin stabilizing factor. *J. Clin. Invest. 66*:1374–1378 (1980).
42. R. P. McDonagh, J. McDonagh, T. E. Petersen, H. C. Thøgersen, K. Skorstengaard, L. Sottrup-Jensen, S. Magnusson, A. Dell, and H. R. Morris, Amino acid sequence of the factor XIIIa acceptor site in bovine plasma fibronectin. *FEBS Lett. 127*:174–178 (1981).
43. H. Hörmann, A. Sternberger, and F. Julek, Interaction of a soluble fibroblast surface protein with fibrin. *Thromb. Haemost. (Suppl.) 63*:379–382 (1978).
44. H. P. Erickson, N. A. Carrell, and J. McDonagh, The structure of fibronectin. *J. Cell. Biol. 91*:673–678 (1981).
45. N. A. Carrell and J. McDonagh, Fibrinogen Chapel Hill II: Defective in reactions with thrombin, factor XIIIa, and plasmin. *Br. J. Haematol. 52*: 35–47 (1982).
46. M. Kaminski and J. McDonagh, Studies on the mechanism of thrombin interaction with fibrin. *J. Biol. Chem. 258*:10530–10535 (1983).

47. C. Skrzynia, H. M. Reisner, and J. McDonagh, Characterization of the catalytic subunit of factor XIII by radioimmunoassay. *Blood 60*:1089-1095 (1982).

48. L. Lorand, K. Campbell-Wilkes, and L. Cooperstein, A filter paper assay for transamidating enzymes using radioactive amine substrates. *Anal. Biochem. 50*:623-631 (1972).

49. M. Hada, M. Katu, S. Ikematsu, M. Fujimaki, and K. Fukutake, Possible cross-linking of factor VIII related antigen to fibrin by factor XIII in delayed coagulation process. *Thromb. Res. 25*:163-168 (1982).

50. T. Tamaki and N. Aoki, Cross-linking of α_2-plasmin inhibitor and fibronectin to fibrin by fibrin-stabilizing factor. *Biochim. Biophys. Acta. 661*: 280-286 (1981).

51. M. Hada, M. Kaminski, and J. McDonagh, unpublished observations,

52. L. A. Shermana nd J. Lee, Fibronectin: Blood turnover in normal animals and during intravascular coagulation. *Blood 60*:558-563 (1982).

53. L. A. Sherman, Fibrinogen turnover: Demonstration of multiple pathways of catabolism. *J. Lab. Clin. Med. 79*:710-723 (1972).

54. J. E. Kaplan and P. W. Snedeker, Maintenance of fibrin solubility by plasma fibronectin. *J. Lab. Clin. Med. 96*:1054-1061 (1980).

55. J. Niewiarowska and C. S. Cierniewski, Inhibitory effect of fibronectin on the fibrin formation. *Thromb. Res. 27*:611-618 (1982).

56. D. K. Galanakis and S. R. Simon, Albumin is a fibrin inhibitor and exhibits marked synergism with fibrinogen against fibrin aggregation. *Blood (Supp. 1) 62*:284a (1983).

57. L. L. Shen, J. Hermans, J. McDonagh, R. P. McDonagh, and M. Carr, Effects of calcium ion and covalent crosslinking on formation and elasticity of fibrin gels. *Thromb. Res. 6*:255-265 (1975).

58. G. W. Kamykowski, D. F. Mosher, L. Lorand, and J. D. Ferry, Modification of shear modulus and creep compliance of fibrin clots by fibronectin. *Biophys. Chem. 13*:25-28 (1981).

59. T. W. Chou, L. V. McIntire, and D. M. Petersen, Importance of plasma fibronectin in determining PFP and PRP clot mechanical properties. *Thromb. Res. 29*:243-248 (1983).

60. L. L. Shen, R. P. McDonagh, J. McDonagh, and J. Hermans, Early events in the plasmin digestion of fibrinogen and fibrin: Effects of plasmin on fibrin polymerization. *J. Biol. Chem. 252*:6184-6189 (1977).

61. F. Grinnell, M. Feld, and D. Minter, Fibroblast adhesion to fibrinogen and fibrin substrata: Requirement for cold-insoluble globulin (plasma fibronectin). *Cell 19*:517-525 (1980).

62. M. P. Bevilacqua, D. Amrani, M. W. Mosesson, and C. Bianco, Receptors for cold-insoluble globulin (plasma fibronectin) on human monocytes. *J. Exp. Med. 153*:42-60 (1981).

63. D. F. Mosher and E. M. Williams, Fibronectin concentration is decreased in plasma of severely ill patients with disseminated intravascular coagulation. *J. Lab. Clin. Med. 91*:729-735 (1978).
64. N. E. Stathakis, A. Fountas, and E. Tsianos, Plasma fibronectin in normal subjects and in various disease states. *J. Clin. Pathol. 34*:504-508 (1981).
65. S. Lopaciuk, K. M. Lovette, J. McDonagh, H. Y. K. Chuang, and R. P. McDonagh, Subcellular distribution of fibrinogen and factor XIII in human blood platelets. *Thromb. Res. 8*:453-465 (1976).
66. J. McDonagh, P. Henriksson, and S. Becker, Identification of factor XIII in monocytes and macrophages. *Circulation 66*:295 (1982).
67. P. J. Birckbichler and M. K. Patterson, Jr., Cellular transglutaminase, growth, and transformation. *Ann. N. Y. Acad. Sci. 312*:354-365 (1978).
68. M. R. Owens and C. D. Cimino, Synthesis of fibronectin by the isolated perfused rat liver. *Blood 59*:1305-1309 (1982).
69. W. A. Scovill, S. J. Annest, T. M. Saba, F. A. Blumenstock, J. C. Newell, H. H. Stratton, and S. R. Powers, Cardiovascular hemodynamics after opsonic α_2-surface binding glycoprotein therapy in injured patients. *Surgery 86*:284-293 (1979).
70. A. B. Robbins, J. E. Doran, A. C. Reese, and A. R. Mansberger, Jr., Clinical response to cold insoluble globulin replacement in a patient with sepsis and thermal injury. *Am. J. Surg. 142*:636-638 (1981).

8

Platelet Function

Mark H. Ginsberg and Edward F. Plow
Scripps Clinic and Research Foundation
La Jolla, California

Gérard A. Marguerie*
Institut de Pathologie Cellulaire
Hôpital de Bicêtre
Bicêtre, France

Blood platelets are anucleate 3 μm diameter cell fragments that circulate at a concentration of 3×10^8 cells per millileter in blood. The primary function of platelets resides in the hemostatic response in which the cells play an essential role in the arrest of bleeding. They perform this function by adhering to subendothelium and connective tissue exposed as a consequence of vessel injury and by adhering to each other to form aggregates that may occlude large defects. These basic reactions—aggregation, adhesion, and subsequent spreading along surfaces—appear to be mediated by at least four distinct large glycoproteins: fibronectin, fibrinogen, thrombospondin, and von Willebrand's factor (VWF). Although this chapter focuses on the role of fibronectin in plate-

Present affiliation: INSERM, Centre d'Etude Nucleaire, Grenoble, France

let function and the possible value of the platelet system as a model for understanding fibronectin interaction with and processing by other cells, we also highlight similarities and differences with the other three members of this "big four."

These four large glycoproteins share a variety of chemical and functional similarities. In addition to being large (M_r = 340,000 to several million) and glycosylated, they each possess intramolecular symmetry. Such symmetry permits a single molecule to exhibit bridging functions between other molecules or cells or between cells and substrata. Further, all are present in platelets, at least three (fibronectin, thrombospondin, and VWF) are made by endothelial cells [1-3] and are incorporated into the matrix, and three (fibronectin, VWF, and fibrinogen) circulate at substantial concentrations in plasma (Table 1).

To participate in adhesive reactions a protein must bind to the cell surface, and each of these proteins binds to platelets (see Refs. 4-7, for example). Second, to bridge cells to substrata, the protein must also interact with the substratum. In blood vessels, relevant substrata may be connective tissue or subendothelium containing collagen, proteoglycans, and noncollagenous proteins or fibrin clots. Certain members of this quartet bind to each of these components. For example, fibrinogen, fibronectin [8], thrombospondin [9], and VWF [10] may bind to fibrin or fibrinogen. It is therefore clear that each of these proteins has chemical and functional properties to serve a role in platelet adhesive function. In

Table 1 Large Adhesive Platelet Glycoproteins

Protein	MW (kilodaltons)	Platelet content (per 10^9 cells)	Plasma content (per ml)
Fibrinogen	340	50 μg[a]	3000 μg[a]
Fibronectin	$\geqslant$450	3 μg[a]	300 μg[a]
von Willebrand's factor	$>10^3$	0.25 u[b]	1 U[b]
Thrombospondin	450	20 μg[a]	20 ng[a]

[a]Based on our measurements in radioimmunoassay and those of others.
[b]Based on Nachman and Jaffe, *J. Exp. Med. 141*:1101 (1975).

this chapter, we consider the following issues concerning fibronectin and platelets: (1) the evidence for a role for fibronectin in
platelet function, (2) the localization of pools of fibronectin involved in platelet function, (3) the processing of the intraplatelet
fibronectin pool, and (4) the identification of and properties of
the platelet binding site(s) for fibronectin. In each case we will
then briefly discuss the similar issues for each of the other members of this group of proteins.

ROLE OF FIBRONECTIN

Adhesion to Substrate

Since fibronectin serves as an attachment protein for several cell
types, binds to platelets, and binds to constituents of the subendothelium and connective tissues, it appears to be a likely condidate
as a platelet attachment protein. Nevertheless, as discussed, this
cell has many potential attachment proteins to choose between. In
addition, since platelets may adhere to certain substrata in the absence of any exogenous adhesive protein [11], the role of endogenous stores of these proteins must be carefully considered. Finally, adhesion as measured by the deposition of cells per unit area
of substratum is critically dependent on shear rate [12] and since
blood vessel diameters vary, no single shear rate can be selected as
the most representative. Thus, measurement of the effect of fibronectin on adhesion becomes a complex and assay-specific task. At-
testing to these complexities has been the variable results reported
on the effect of plasma fibronectin on platelet adhesion to collagenous substrate. Two groups found little effect of plasma fibronectin on the number of platelets adherent to collagenous substrata [13, 14]; a third reported substantial enhancement [15]. A
fourth investigator reported that plasma fibronectin slightly inhibited platelet adhesion to fibrillar collagen in a gel filtration assay [16]. These potential controversies obviously may relate to
the previous discussion and to variable contributions of endogenous secreted adhesive proteins and aggregation in the different assays used. Nevertheless, there is clear agreement among three
groups [13-15] that plasma fibronectin promotes platelet spreading on collagenous surfaces. Since such spreading probably signi-
fies an increase in the area of cell-substratum contact, fibronectin

is likely to promote increased strength of cell attachment. Thus, the clear demonstration of the ability of fibronectin to promote platelet binding to collagenous substrata will probably require studies at a variety of shear rates with careful consideration of the possible role of endogenous fibronectin stores, as was required to demonstrate such an effect for VWF on platelet vessel wall interactions [17]. A preliminary report of studies of this type has appeared [18].

Aggregation

Platelets require a plasma cofactor to aggregate in response to ADP. It has been clearly established [19] that fibronectin does not serve this function. In addition, platelet aggregation in response to ADP or thrombin proceeds normally in fibronectin-depleted plasma [20]. Thus, plasma fibronectin does not appear to play an important role in aggregation responses to these agonists. In contrast, Arneson et al. [21] reported a patient with Ehlers-Danlos syndrome whose collagen-induced platelet aggregation was abnormal; the abnormality could be corrected by addition of purified plasma fibronectin back to the patient's plasma. This observation is difficult to reconcile with data indicating that collagen-induced platelet aggregation is normal [20] or increased [22] in fibronectin-depleted plasma. Taken together, these data do not provide a compelling case for a role for plasma fibronectin in platelet aggregation. The possibility that platelet fibronectin plays a contributory role in aggregation occurring in secreting cells cannot be excluded, and Santoro [23] has recently reported that fibronectin may inhibit thrombin-induced aggregation. He has interpreted his findings as suggesting a role for platelet fibronectin in thrombin-induced platelet-platelet interactions.

Fibronectin and the Collagen Receptor

Bensusan and coworkers [24] observed that a molecule of similar size to fibronectin remained bound when collagen-adherent platelets were sonicated, that fibronectin inhibited collagen-induced aggregation, and that anti-fibronectin stimulated platelet secretion. Based on these observations they proposed that cell surface fibronectin acts as a collagen receptor on platelets. Unfortunately, the surface localization of the putative fibronectin was not established in that study, and it is now clear that little if any fibronectin is present on the surface of resting platelets [25, 26]. Further, digestion of intact platelets with proteases known to cleave fibronectin [27] or anti-fibronectin antibodies [16] has little effect on platelet stimulation by collagen. Further studies by the Western Reserve Group [28] have established that the ability of the anti-fibronectin antibodies to stimulate platelets may reside in the formation of immune complexes rather than by cross-linking a membrane protein. Thus, although the hypothesis that cell surface fibronectin acts as a receptor for a platelet agonist is exciting, its absence from the surface of resting cells, coupled with the protease and antibody data just cited, make the possibility appear unlikely.

LOCALIZATION OF FIBRONECTIN

As is noted elsewhere in this volume, fibronectin is a constituent of connective tissue matrices and a plasma protein (concentration ~600 nM) and is synthesized by cultured endothelial cells [2]. Thrombospondin is made by endothelial cells [3] and fibroblasts [29] but is present at very low levels in plasma [30]; VWF appears to be made in endothelium [1] and is a plasma protein. Each of these three proteins is also found in association with cell matrices [29, 31]. In contrast, fibrinogen is made primarily in the liver [32] and is a plasma protein that does not appear to be a normal matrix constituent. In this section we focus on the evidence for platelet pools of each of these proteins and their localization in resting platelets.

Platelet Fibronectin

The presence of fibronectin antigen in platelets was first suggested by the observation of a precipitin line when a platelet extract was reacted with an antiserum to cold-insoluble globulin [33]. Subsequently, platelet fibronectin antigen was detected by radioimmunoassay [34] and electroimmunoassay [35], and levels of ~3 μg per 10^{-9} cells were found by both methods. Since platelet fibronectin antigen was immunochemically identical to the plasma form and approximately 300 μg/ml is present in normal human plasma, it was important to exclude plasma contamination as the source of the platelet antigen. This was done by use of exogenous radiolabeled plasma fibronectin as a control for plasma contamination of suspensions of washed platelets and direct immunofluorescent demonstration of the platelet fibronectin antigen. The latter control also served to exclude the possibility that the measured fibronectin antigen was associated with a contaminating cell type [34]. It should be noted that platelet fibronectin antigen also shares the property of binding to gelatin [34] but has not been further characterized with respect to its structure.

Although initial workers proposed [24] that platelet fibronectin might be a cell surface molecule, later studies bring this concept into question. Cell surface staining for fibronectin has not been consistently observed in resting platelets by immunofluorescence [13, 25]. Second, lactoperoxidase-catalyzed iodination of intact platelets does not result in labeling of any species of the size of fibronectin [6]. Third, on subcellular fractionation, the bulk of platelet fibronectin antigen associates with α granule rather than membrane fractions [35]. Thus, it appears that the bulk of platelet fibronectin antigen is not on the cell surface. Moreover (see below), since at least two mechanisms exist for the appearance of fibronectin on the surface of activated platelets, the finding of fibronectin on the surface of resting cells may be due to low-level activation. Indeed, we have found that the binding of radiolabeled affinity-purified F(ab)'2 anti-fibronectin to "resting" platelets could be completely inhibited when the cells were washed in the presence of an inhibitor of their activation [36]. This caveat also pertains to the detection of other members of the big four on the surface of resting cells.

In contrast to the lack of convincing data concerning the presence of fibronectin on the surface of resting cells, there is compelling evidence for storage of platelet fibronectin in organelles termed α granules. First, on subcellular fractionation, platelet fibronectin cosediments with authentic α-granule markers (see Fig. 1) [35]. Second, platelets congenitally deficient in α granules have reduced fibronectin content [37, 38]. Third, platelet fibronectin is actively secreted along with other α-granule constituents [25, 35]. Finally, by immunofluoresence, platelet fibronectin has a similar intracellular localization to other α-granule constituents [25, 39, 40] and in frozen thin sections has recently been localized to the α granule [41].

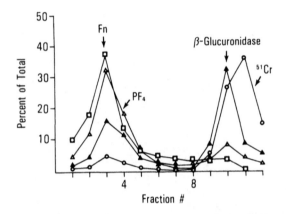

Figure 1 Subcellular localization of platelet fibronectin antigen. Resting platelets (10^9 ml^{-1}) were homogenized by nitrogen cavitation (1200 psi, 15 min equilibration at 4° C). The homogenate was layered over a 20–60% sucrose solution and centrifuged at 36,000 RPM for 120 min in an SW41 rotor at 4° C. Fractions of 1 ml (fraction 12 = 0.5 ml) were collected, 1% Triton X-100 was added, and content of ^{51}Cr (o———o) assayed by radioactivity, platelet factor 4(△———△), β-thromboglobulin (not shown), and fibronectin (□———□) were assayed in radioimmunoassay, protein was assayed by the Lowry method (not shown), and β-glucuronidase (▲———▲) was measured by hydrolysis of phenolphthalein β-glucuronide. Protein showed a similar sedimentation pattern to ^{51}Cr. β-Thromboglobulin was similar to platelet factor 4, > 80% of each marker was recovered from the gradient.

Localization of Other Adhesive Proteins

In the case of the other three adhesive proteins, evidence exists that they, too, are present in α granules, including subcellular fractionation [35, 42, 43], deficiency in α-granule–deficient platelets [37, 38, 44], secretion [45], immunofluorescence [39, 40, 46], and immunoelectron microscopy [41, 47]. Further, each of the adhesive proteins appears to be present in the same population of α granules (Figs. 2 and 3) [40].

Surface Expression

If these platelet glycoproteins are to serve an adhesive function they must reach the cell surface. There is compelling evidence that

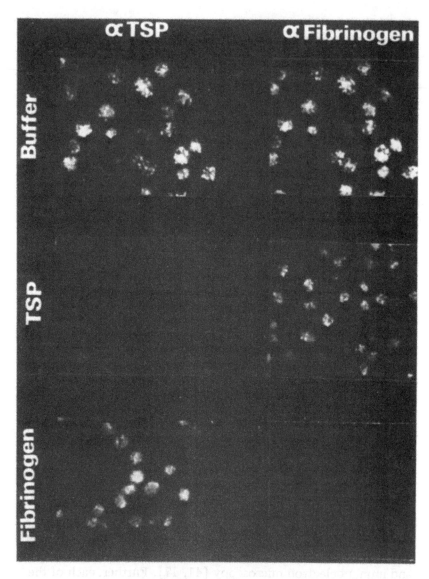

Figure 2 Colocalization of thrombospondin and fibrinogen in resting platelets. Resting permeable platelets were stained with a mixture of rabbit anti-
thrombospondin and rhodaminated goat anti-fibrinogen, and counterstained
with fluoresceinated goat F(ab′)2 anti-rabbit IgG. Antibodies absorbed with
buffer demonstrated punctate intracellular fluorescent staining, which exactly
colocalized for both ligands. Incubation of the antibody mixture prior to
staining with either purified thrombospondin or purified fibrinogen only
blocked fluorescence when the antibody was absorbed with its parent antigen. (X1800) (From Ref. 40.)

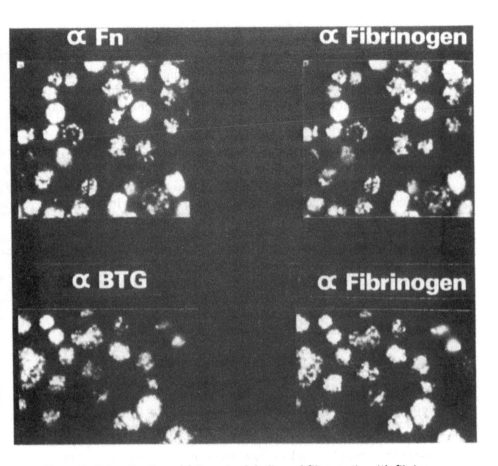

Figure 3 Colocalization of β-thromboglobulin and fibronectin with fibrinogen in resting platelets. Resting permeable platelets were stained as in Figure
2 with mixtures of rabbit anti-fibronectin or β-thromboglobulin and rhodaminated goat anti-fibrinogen. Fibronectin and β-thromboglobulin exactly colocalized with fibrinogen at this resolution. (X1700) (From Ref. 40.)

each of these proteins is expressed on the surface of activated
platelets [6, 26, 48, 49]. In the case of fibronectin, this has been
established by immunofluorescence and by binding of affinity-
purified radiolabeled antibody [26]. Two mechanisms could account for these observations. First, secreted proteins could rebind

to the cell surface, or second, they may be prebound to granule membranes and be expressed on the cell surface as a sticky patch as a direct consequence of the secretory granule-plasma membrane fusion event. In the case of fibronectin, the initial surface expression is probably too rapid to be accounted for by rebinding [36], suggesting that at early time points the latter mechanism predominates. Moreover, we [40] have recently observed initial colocalization of all four adhesive proteins as they reach the platelet surface. Such a rapid localized expression of adhesive proteins on the cell surface has interesting implications. First, it may partially account for the close kinetic relationship between platelet secretion and secondary aggregation [50]. Second, if as occurs in other cell types [51, 52], secretion along an activating surface may be polarized toward that surface. Polarized expression of adhesive proteins in platelets adherent to a vessel wall would result in these proteins being present only on the vessel side, not the luminal side. Thus, the luminal surface of such adherent platelets might be relatively less thrombogenic than the adherent side so that aggregation would not be a necessary concomitant of the adhesive event.

For such a mechanism of polarized rapid secretion to occur, several requirements should be met. The proteins should be prebound to receptors on the inner surface of the α-granule membrane. Recently, Gogstad et al. [53] reported that a molecule of similar size to thrombospondin TSP coisolated with a particulate fraction of ruptured α granules. Second, the same group [54] presented evidence to suggest that membrane glycoprotein (GP) IIb/III (a potential fibrinogen receptor; see, for example, Ref. 55) was present in α granules, and we have recently [56] observed an intracellular pool of GP IIb/III by immunofluorescence. Thus, there is evidence for an α-granule pool of potential receptors for at least one of these proteins. A third requirement of this hypothesis is that the α granules fuse with the plasma membrane. Although secretion of α-granule constituents is nonlytic, fusion of these granules with the platelet plasma membrane has not been documented, suggesting that they may not secrete by a simple exocytotic mechanism. Behnke [57] and White [58] provided evidence that platelets contain an open canalicular system in free communication with the extracellular space, leading White [58] to propose that secretion is into this system. White and Clawson [59] have found that the canaliculi are an interconnected meshwork unlikely to be evaginated during secretion. Thus, the hypothesized insertion of adhesive proteins would occur in the depth of the canalicular system, a site unlikely to be accessible to substrata or other cells. In our studies of release of α-granule constituents using platelet factor 4 as a model, we found that it was consolidated into larger pools prior to secretion (Fig. 4), and a similar redistribution has been observed for the four adhesive proteins [40]. These pools appeared as vacuolar structures in transmission electron microscopy and were closed to probes of the extracellular space [60]. These data are compatible with α-granule secretion into an intermittently open canalicular system, as suggested by Zucker-Franklin [61], but are also compatible with a compound exocytotic mechanism. In the latter case, we propose that α-granule constituent release occurs by initial fusion of these granules with each other and/or another intracellular compartment to form a morphologically distinct compound granule followed by compound granule fusion with the plasma membrane. Possible instances of the latter have been observed in both transmission electron microscopy [62] and living cells [63]. Thus, we hypothesize that these adhesive proteins may be transported to the platelet surface by a compound exocytotic mechanism resulting in their rapid expression at the cell surface initially in localized, potentially polarized patches.

Once on the platelet surface, fibronectin undergoes a time-dependent redistribution to a more compact distribution. In immunofluorescence, this redistribution bears a passing resemblance to the "capping" that occurs in other cell types [26]. Since capping may involve transmembrane interactions between cell surface and cytoskeletal proteins [64], the possibility exists that cell surface fibronectin or other adhesive proteins enter into such transmembrane associations. Phillips et al. [65] reported that a portion of platelet membrane glycoprotein IIb and III associated with the Triton-insoluble residue of thrombin-aggregated platelets, although subsequent work [66] has raised the disturbing possibility that the retained glycoproteins were merely trapped within undissolved phospholipid bilayers. In addition, we [67] rigorously established that these membrane glycoproteins were physically associated with the cytoskeleton of concanavalin A-activated platelets. First, these glycoproteins cosedimented with the Triton-insoluble "cyto-

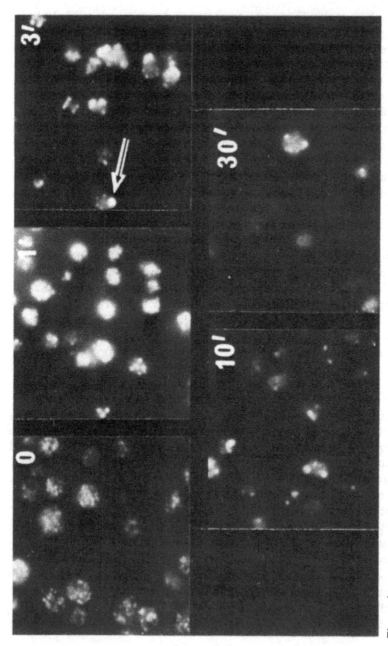

Figure 4

skeleton" in concanavalin A-activated cells, and this is unassociated with an increase in retained lipid phosphorus [66]. Second, this cosedimentation required that the concanavalin A bind to intact cells and the glycoproteins could be solubilized by depolymerization of F-actin by DNase I. Third, the cytoskeleton-associated concanavalin A was demonstrated to be bound to the periphery of the filamentous cytoskeleton in frozen thin section of these Triton residues. Finally, surface-bound conconavalin A could be induced to "cap" by a thrombin-stimulated redistribution of the underlying cytoskeleton, providing evidence of a transmembrane interaction in intact cells. Since each of these adhesive proteins has intramolecular symmetry, the possibility that they, like concanavalin A, act as multivalent ligands seems likely. Thus, these glycoproteins may also enter into such transmembrane associations, as suggested by the data discussed in a later section.

FIBRONECTIN BINDING TO PLATELETS

There is considerable interest in the definition of cellular fibronectin binding sites. In this regard the platelet may prove valuable in that the cell saturably binds plasma fibronectin when stimulated with appropriate agonists.

Requirements for Expression of Platelet Fibronectin Binding

Incubation of ^{125}I-labeled plasma fibronectin with thrombin-stimulated suspensions of washed human platelets results in time-dependent binding of fibronectin to the cells [4]. Binding requires ~20 min to reach a maximum and does not occur in resting cells.

Figure 4 Visualization of thrombin effects on platelet factor 4 (PF4) distribution. A suspension of washed platelets containing 5×10^8 cells ml^{-1} was stimulated at $22°C$ with 0.5 U/ml human thrombin. At the indicated times, aliquots were removed, fixed, and stained in the permeable state for PF4. Numbers in the upper right-hand corner refer to time in minutes after thrombin addition. Note that the initially fine punctate PF4 distribution appears to consolidate into a few large masses of fluorescence, which are seen at the cell periphery. Note the markedly reduced number of masses at 30 min. Similar numbers of cells were photographed for each field. (X1800) (From Ref. 60.)

The specificity of this reaction was documented by the ability of fibronectin but not unrelated proteins or fibronectin-depleted serum to inhibit binding. The ability of fibronectin to inhibit binding indicates a saturable process, and by Scratchard analysis 120,000 molecules per cell are bound at saturation. Half-saturation of cellular fibronectin binding sites is achieved at 3×10^{-7} M fibronectin. Binding occurs in the presence of 5 mM MgEDTA, suggesting a lack of an absolute calcium requirement, but was inhibited in the presence of 5mM EDTA, indicating a requirement for divalent cations.

Identification of Platelets Congenitally Deficient in Fibronectin Binding

Glanzmann's thrombasthenia is an autosomally inherited recessive severe bleeding disorder that results from an intrinsic platelet defect. In vitro abnormalities of individuals with this disorder include reduced or absent platelet aggregation in response to collagen, thrombin, and ADP, reduced clot retraction, and abnormal platelet spreading on glass surfaces [68]. In addition, patients with this disease generally [68] but not always [69] are deficient in a [70] specific platelet membrane glycoprotein, GPIIb/III. Platelets from patients with thrombasthenia bind very little plasma fibronectin when stimulated with thrombin (Fig. 5) [71, 72]. This failure to bind plasma fibronectin is not due to a lack of platelet stimulation by thrombin, or to a soluble inhibitor of fibronectin binding, but rather to a reduced number of fibronectin binding sites [71]. Thus, an autosomal recessive disorder is associated with reduced fibroblast retraction of fibrin clots [73] and collagen gels [74].

Role of Secreted Proteins in Fibronectin Binding

As noted, thrombin induces fibronectin binding sites on platelets and immunofluorescence suggests that collagen may do so as well [13]. Both these stimuli induce secretion of platelet α-granule constituents and surface expression of the adhesive proteins. In contrast, in our assay conditions neither epinephrine nor ADP is a potent inducer of fibronectin binding sites, although they do induce fibrinogen receptors [4]. This raises the possibility that a

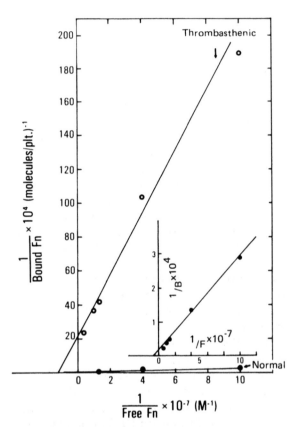

Figure 5 Reduced binding of plasma fibronectin to thrombasthenic platelets. Varying concentrations of ^{125}I-labeled plasma fibronectin (10^{-8}-10^{-6}
M) were incubated for 30 min with thrombin-stimulated normal (●——●)
or thrombasthenic (O——O) platelets and bound ligand was measured by
centrifugation through 20% sucrose. Data are shown as double reciprocal
plots. Inset: replot of normal data with expanded ordinate. Sites per cell from
Y intercept: normal = 80,000, thrombasthenic = 500; apparent K_d from X
intercept: normal = 2.2×10^{-7} M, thrombasthenic = 1.0×10^{-7} M. (From
Ref. 71.)

newly surface expressed α-granule constituent mediates fibronectin binding to platelets.

One candidate α-granule protein is platelet fibrinogen, which is present at a level of approximately 100,000 molecules per cell. In addition, thrombasthenic platelets often have a reduced fibrinogen content and fail to express fibrinogen receptors in response to ADP [75], raising the possibility that a fibrinogen-related defect may account for the defect in fibronectin binding. Further, fibrin interacts with fibronectin, although the noncovalent interaction is of low affinity at 37°C, the incubation temperature of our binding assays [8]. Nevertheless, platelet fibrinogen is not absolutely required for fibronectin binding since afibrinogenemic platelets, containing less than 2% the normal fibrinogen content, bind fibronectin [71]. There may, however, be a fibrinogen-dependent pathway of fibronectin binding to platelets, since a portion of platelet-bound fibronectin appears to be covalently cross-linked in a manner compatible with factor XIII-mediated cross-linking to fibrin [4]. Moreover, this cross-linking has an absolute calcium dependence, in contrast to the noncovalent association of fibronectin with platelets [76], and can be reconstituted in vitro with exogenous fibrin bound to platelet fibrinogen receptors.

A second candidate α-granule protein is thrombospondin, which is present at approximately 40,000 molecules per cell. Since thrombospondin is a trimer [43], this means that thrombospondin subunits could accommodate fibronectin binding at a 1:1 stoichiometry. Thrombospondin binds to fibronectin [9, 77] and is synthesized by fibroblasts [29] and endothelial cells [3], appears to be a matrix protein [29], and may therefore support fibronectin interaction in other cell types as well. Lehav et al. [77] found that, when platelets react with surface-bound fibronectin derivatized with N-succinimidyl-3-[2-nitro-4-aziodophenyl)-2-amino-ethldithiolproprionate, photoactivation results in cross-linking to thrombospondin. Nevertheless, thrombospondin is unlikely to serve as a major univalent fibronectin receptor on thrombin-activated platelets because gray platelets containing less than 5% of normal thrombospondin bind plasma fibronectin normally in response to thrombin [38]. Moreover, since these platelets are deficient in other α-granule constituents as well, such constituents are unlikely to be required for thrombin-induced fibronectin binding.

Association of Fibronectin with the Platelet Cytoskeleton

As noted, multivalent ligands may trigger association of membrane proteins with the cytoskeleton, resulting in loss of Triton solubility of the platelet-bound ligand. The data summarized in Table 2 suggest that fibronectin is capable of participating in such transmembrane associations with the platelet cytoskeleton. ^{125}I-labeled plasma fibronectin was added to thrombin-stimulated or resting platelets, and as anticipated, significant surface binding occurred only to the stimulated cells. These cells were then extracted with Triton X-100 as described [67]. After two washings, the Triton residue-associated radioactivity was not further decreased by washing and contained a substantial portion of the bound radioactivity of the stimulated cells. In Table 2, this represented 65% of bound radioactivity and ranged from 25 to 95% in five separate experiments. When specific fibronectin binding was inhibited by excess cold ligand or EDTA, or when Triton was added to the cells prior to fibronectin, the Triton residue-associated radioactivity was reduced to negligible levels. These experiments establish reaction specificity and the requirement for fibronectin binding to the platelet surface. In addition, the Triton insolubilization of the labeled fibronectin depended on actin sedimentability since DNase I, which solubilizes F-actin from these platelet cytoskeletons [67] decreased fibronectin recovery in the Triton residue. Polyacrylamide gel analysis confirmed (not shown) that the effect of DNase occurred in the absence of proteolysis of the fibronectin or of the cytoskeletal proteins.

In parallel experiments, ^{125}I-labeled fibrinogen also associated with the Triton residue of thrombin-stimulated platelets. This association also occurred with resting cells and in cells prelysed with Triton. In addition, it was not blocked by addition of EDTA. Thus, the capacity of fibronectin to coisolate with the Triton residue requires cell stimulation, intact cells, and divalent ions that would support specific surface binding. In contrast, in the case of fibrinogen, although association with the Triton residue occurred, a requirement for specific surface binding could not be demonstrated.

Table 2 Interaction of Fibronectin with the Platelet Cytoskeleton[a]

	Fibronectin bound	
Condition	Molecules/platelet	Molecules in cytoskeleton
Platelets	2,400	2,200
Platelets + thrombin	64,000	37,600
Platelets + thrombin + EDTA	3,200	2,800
DNase I treatment	—	13,200

[a] ^{125}I-labeled fibronectin at 0.3 μM was added to nonstimulated platelets or thrombin-stimulated platelets (treated for 5 min with 2 u/ml thrombin and then excess hirudin added) in Tyrode's-albumin buffer containing 1 mM calcium or 2 mM EDTA. After 30 min at 37° C without stirring, a portion of the cells were removed and the fibronectin bound per platelet was measured by centrifugation through sucrose (Plow and Ginsberg, 1981). Triton X-100 (2%) containing 2 mM EDTA was added to the rest of the cells. After 30 min, the samples were centrifuged to recover the cytoskeletal pellets, and these were washed two additional times to determine the fibronectin molecules associated with the cytoskeleton. In addition, DNase I (1 mg/ml containing 0.2 mM PMSF) was added to the Triton extract of thrombin-stimulated platelets (with calcium), and the fibronectin associated with the cytoskeleton was determined as indicated above.

Relationship to Other Adhesive Protein Binding Sites

As noted, all four of these adhesive proteins bind to thrombin-activated platelets [4-6, 78]. In three cases (VWF, fibrinogen, and fibronectin), binding is inhibited by ADP scavengers (our unpublished observatons) [7, 79] and in every case is dependent on divalent cations. Glanzmann's thrombasthenic platelets are deficient in binding of at least three of these proteins [72, 75, 80], and monoclonal antiplatelet antibodies, which appear to react with the GPIIb/III complex, coordinately inhibit the binding of all four ligands [81]. Taken together, these data suggest some relationship in the platelet binding sites for each of these proteins.

1. Does one of these proteins serve as the binding site for the other three? One possible relationship among these binding sites is that one of these proteins serves as the primary binding site for

the others, particularly in view of the evidence of extensive interactions among this group. Since ADP-stimulated fibrinogen binding occurs in the absence of platelet secretion or evident change in
the surface protein composition [5, 82] it is unlikely that another
adhesive protein serves as a primary fibrinogen binding site. Conversely, since fibrinogen or fibrin bind to the other three proteins
[8-10], fibrinogen might serve as such a common binding site.
Moreover, as noted, such a mechanism would account for the multiple deficiences in thrombasthenic platelets. This appears unlikely
since fibronectin [71] and VWF [80] bind to thrombin-stimulated afibrinogenemic platelets. In addition, in the case of fibronectin, we also know [38] that other constituents of α granules
are unlikely to be essential since gray platelets bind fibronectin.
Thus it appears that a single one of these proteins cannot serve as
the primary binding site for all of the other three. Conversely, it is
possible that interactions among these proteins may contribute to
interactions with platelets. For example, although fibrongien is
not required for fibronectin binding, a portion of platelet-bound
fibronectin appears to be covalently cross-linked to platelet fibronectin [4].

   2. Do these proteins bind to the same site on the platelet surface? As noted, there are remarkable similarities in binding site induction requirements among the big four, and there appears to be
a coordinate reduction in site expression for at least three of them
in Glanzmann's thrombasthenia. Since the latter disease is associated with a specific membrane protein deficit, the possibility
arises that the membrane protein (GPIIb/III) serves as the receptor
for all four proteins. It is clear that fibrinogen receptors may be
induced without expression of fibronectin binding activity and
that there are approximately 2.5 times more fibronectin than fibrinogen binding sites [4], indicating that there is not complete
concordance between the two sites. Second, it has been reported
that fibronectin does not compete for VWF binding sites [7], although fibrinogen appears to do so [83]. Similarly, fibrinogen inhibits fibronectin binding, but fibronectin has little effect on fibrinogen binding [76]. Since fibrinogen and fibronectin bind with
similar affinities and initial rates, these two points militate against
a simple single common binding site but may indicate that these
proteins bind to platelets through multiple subsites, some of which

they may share. Indeed, synthetic peptides from the γ chain of fibrinogen of $M_r < 2000$ coordinately inhibit both fibrinogen and VWF binding [84, 85], as well as fibronectin binding [85] to thrombin-activated platelets. Deficiency of a single essential common subsite could readily account for the multiple deficiencies in thrombasthenia. This hypothesis could probably be further explored by a detailed analysis of the steric relationship among binding sites for these ligands using the ligands themselves, monoclonal antibodies against GPIIb/III, and if available, other fragments of the ligands containing platelet reactive sites, in competitive binding assays.

CONCLUDING REMARKS

The discovery, characterization, chemistry, and biology of adhesive proteins has provided fruitful insights into a broad group of problems in cell biology. Platelets are anucleate cell fragments programmed to express adhesive activities in response to vessel injury. As indicated in this chapter, there are at least four candiate, well-characterized adhesive proteins that may subserve these activities and considerable data concerning the identity of and regulation of receptors for these proteins. This "cell" thus offers a useful model for the study of cell adhesion and cell cohesion. Moreover, due to the existence of deficiency diseases, we know that fibrinogen (MW = 340,000) and VWF (MW > 1,000,000) play essential physiological roles in platelet function. In contrast, in the case of thrombospondin (MW = 450,000) and fibronectin (MW = 450,000), although there are good reasons to believe they function in hemostasis, the lack of selective deficiency diseases makes their function less certain.

ACKNOWLEDGMENTS

The authors are deeply indebted to Jane Forsyth for the technical conduct of the majority of our experiments described here. We also acknowledge the artwork of Betsy Cargo and the expert secretarial assistance of N. L. S. Kaypro, II.

Supported by Grants #HL 16411, HL 28235, and AM 27214 from the NIH. M. H. G. is recipient of RCDA #AM 00720.

REFERENCES

1. E. A. Jaffe, L. W. Hoyer, and R. L. Nachman, Synthesis of antihemo-
   philic factor antigen by cultured human endothelial cells. *J. Clin. Invest.*
   52: 2757-2764 (1973).
2. E. A. Jaffe and D. F. Mosher, Synthesis of fibronectin by cultured human endothelial cells. *J. Exp. Med. 147*:1779-1791 (1978).
3. D. F. Mosher, M. J. Doyle, and E. A. Jaffe, Synthesis and secretion of
   thrombospondin by cultured human endothelial cells. *J. Cell Biol. 93*:
   343-348 (1982).
4. E. F. Plow and M. H. Ginsberg, Specific and saturable binding of plasma
   fibronectin to thrombin-stimulated human platelets. *J. Biol. Chem. 256*:
   9477-9482 (1981).
5. G. A. Marguerie, E. F. Plow, and T. S. Edgington, Human platelets possess
   an inducible and saturable receptor specific for fibrinogen. *J. Biol. Chem.*
   254:5357-5363 (1979).
6. D. R. Phillips, L. K. Jennings, and H. R. Prasanna, Ca^{2+}-mediated association of glycoprotein G (thrombin-sensitive protein, thrombospondin)
   with human platelets. *J. Biol. Chem. 255*:11629-11632 (1980).
7. T. Fujimoto, S. Ohara, and J. Hawiger, Thrombin-induced exposure and
   prostacyclin inhibition of the receptor for Factor VIII/von Willebrand
   Factor on human platelets. *J. Clin. Invest. 69*:1212-1222 (1982).
8. E. Ruoslahti and A. Vaheri, Interaction of soluble fibroblast surface antigen with fibrinogen and fibrin. Identity with cold insoluble globulin
   of human plasma. *J. Exp. Med. 141*:497-501 (1975).
9. L. L. K. Leung and R. L. Nachman, Complex formation of platelet
   thrombospondin with fibrinogen. *J. Clin. Invest. 70*:542-549 (1982).
10. D. L. Amrani, M. W. Mosesson, and L. W. Hoyer, Distribution of plasma
    fibronectin (cold-insoluble globulin) and components of the factor VIII
    complex after heparin-induced precipitation of plasma. *Blood 59*:657–
    663 (1982).
11. P. J. Shadle and S. H. Barondes, Adhesion of human platelets to im-
    mobilized trimeric collagen. *J. Cell Biol. 95*:361-365 (1982).
12. V. Turitto and H. Baumgartner, in *Measurements of Platelet Function*
    (L. Harker and T. Zimmerman, eds.). Churchill Livingstone, New York,
    1983, 46-63.
13. R. O. Hynes, I. U. Ali, A. T. Destree, V. Mautner, M. E. Perkins, D. R.
    Senger, D. D. Wagner, and K. K. Smith, A large glycoprotein lost from
    the surfaces of transformed cells. *Ann. N. Y. Acad. Sci. 312*:317-3432
    (1978).
14. F. Grinnell and D. G. Hays, Cell adhesion and spreading factor. Similar-
    ity to cold insoluble globulin in human serum. *Exp. Cell Res. 115*:221-
    229 (1978).

15. E. I. Chazov, A. V. Alexeev, A. S. Antonov, V. E. Koteliansky, V. L. Leytin, E. V. Lyubimova, V. S. Repin, D. D. Sviridav, V. P. Torchilin, and V. N. Smirnov, Endothelial cell culture on fibrillar collagen: Model to study platelet adhesion and liposome targeting to intercellular collagen matrix. *Proc. Natl. Acad. Sci. USA 78*:5603-5607 (1981).

16. S. A. Santoro and L. W. Cunningham, Fibronectin and the multiple interaction model for platelet-collagen adhesion. *Proc. Natl. Acad. Sci. USA 76*:2644-2648 (1979).

17. H. R. Baumgartner, T. B. Tschopp, and D. Meyer, Shear rate dependent inhibition of platelet adhesion and aggregation on collagenous surfaces by antibodies to human factor VIII/von Willebrand factor. *Br. J. Haematol. 44*:127-139 (1980).

18. W. Houdijk, K. Sakariassan and J. Sixma, *Thromb. Hemo. 50*:127a (1980).

19. M. B. Zucker, M. W. Mosesson, M. J. Broekman, and K. L. Kaplan, Release of platelet fibronectin (cold-insoluble globulin) from alpha granules induced by thrombin or collagen; lack of requirement for plasma fibronectin in ADP-induced platelet aggregation. *Blood 54*:8-12 (1979).

20. I. Cohen, E. V. Potter, T. Glaser, R. Entwistle, L. Davis, J. Chediak, and B. Anderson, Fibronectin in von Willebrand's disease and thrombasthenia: Role in platelet aggregation. *J. Lab. Clin. Med. 97*:134-140 (1981).

21. M. A. Arneson, D. E. Hammerschmidt, L. T. Furcht, and R. A. King, A new form of Ehlers-Danlos syndrome. Fibronectin corrects defective platelet function. *JAMA 244*:144-147 (1981).

22. D. G. Moon and J. E. Kaplan, Plasma fibronectin inhibition of collagen and ADP-induced plately aggregation. *Fed. Proc. 40*:3060 (1981).

23. S. A. Santoro, Inibition of platelet aggregation by fibronectin. *Biochem. Biophys. Res. Commun. 116*:135-140 (1983).

24. H. B. Bensusan, T. L. Koh, K. G. Henry, B. A. Murray, and L. A. Culp, Evidence that fibronectin is the collagen receptor on platelet membranes. *Proc. Natl. Acad. Sci. USA 75*:5864-5968 (1978).

25. M. H. Ginsberg, R. G. Painter, C. Birdwell, and E. F. Plow, The detection, immunofluorescent localization, and thrombin induced release of human platelet-associated fibronectin antigen. *J. Supramol. Struct. 11*:167-173 (1979).

26. M. H. Ginsberg, R. G. Painter, J. Forsyth, C. Birdwell, and E. F. Plow, Thrombin increases expression of fibronectin antigen on the platelet surface. *Proc. Natl. Acad. Sci. USA 77*:1049 (1980).

27. B. Jaques and M. H. Ginsberg, The role of cell surface proteins in platelet stimulation by monosodium urate crystals. *Arthritis Rheum. 25*:508-520 (1982).

28. D. Holderbaum, L. A. Culp, H. B. Bensusan, and H. Gershman, Platelet stimulation by antifibronectin antibodies requires the Fc region of antibody. *Proc. Natl. Acad. Sci. USA* 79:6537-6540 (1982).

29. G. J. Raugi, S. M. Mumby, D. Abbott-Brown, and P. Bornstein, Thrombospondin: Synthesis and secretion by cells in culture. *J. Cell Biol.* 95: 351-354 (1982).

30. S. D. Saglio and H. S. Slayter, Use of a radioimmunoassay to quantify thrombospondin. *Blood* 59:162-166 (1982).

31. D. D. Wagner, J. B. Olmsted, and V. J. Marder, Immunolocalization of von Willebrand protein in Weibel-Palade bodies of human endothelial cells. *J. Cell Biol.* 95:355-360 (1982).

32. L. L. Miler and W. F. Bale, Synthesis of all plasma protein fraction except gamma globulins by the liver. *J. Exp. Med.* 99:125-140 (1954).

33. M. W. Mosesson and R. A. Umfleet, The cold-insoluble globulin of human plasma. I. Purification, primary characterization, and relationship to fibrinogen and other cold-insoluble fraction components. *J. Biol. Chem.* 245:5728-5736 (1970).

34. E. F. Plow, C. Birdwell, and M. H. Ginsberg, Identification and quantitation of platelet-associated fibronectin antigen. *J. Clin. Invest.* 63:540-543 (1979).

35. M. B. Zucker, M. J. Broekman, and K. L. Kaplan, Factor VIII-realted antigen in human blood platelets. Localication and release by thrombin and collagen. *J. Lab. Clin. Med.* 94:675-682 (1979).

36. M. H. Ginsberg and E. F. Plow, Fibronectin expression on the platelet surface occurs in concert with secretion. *J. Supramol. Struct.* 18:91-98 (1981).

37. A. T. Nurden, T. J. Kunicki, D. Dupuis, C. Soria, and J. P. Caen, Specific protein and glycoprotein deficiencies in platelets isolated from two patients with the gray platelet syndrome. *Blood* 59:709-718 (1982).

38. M. H. Ginsberg, J. Durnell, J. G. White, and E. F. Plow, Binding of fibronectin to alpha granule deficient platelets. *J. Cell Biol.* 97:571-574 (1983).

39. J. C. Giddings, L. R. Brookes, F. Piovella, and A. L. Bloom, Immunohistological comparison of platelet factor 4 (PF4), fibronectin (Fn) and factor VIII related antigen (VIIIR:Ag) in human platelet granules. *Br. J. Haematol.* 52:79-89 (1982).

40. J. Wencel-Drake, E. F. Plow, T. S. Zimmerman, R. G. Painter, and M. H. Ginsberg, Immunofluorescent localization of adhesive glycoproteins in resting and thrombin stimulated platelets. *Am. J. Pathol.* in press.

41. J. Wencel-Drake, E. F. Plow, R. G. Painter, T. S. Zimmerman, and M. H. Ginsberg, Ultrastructural localization of human platelet thrombospondin, fibrinogen, fibronectin, and von Willebrand factor in frozen thin section. *Blood (Suppl. 1)* 69:271a (1983).

172   GINSBERG, PLOW, AND MARGUERIE

42. M. J. Broekman, R. I. Handin, and P. Cohen, Distribution of fibrinogen and platelet factors 4 and XIII in subcellular fractions of human platelets. *Br. J. Haematol. 31*:51-55 (1975).
43. J. W. Lawler, H. S. Slayter, and J. E. Coligan, Isolation and characterization of a high molecular weight glycoprotein from human blood platelets. *J. Biol. Chem. 253*:8609-8616 (1978).
44. J. M. Gerrard, D. R. Phillips, G. H. R. Rao, E. F. Plow, D. A. Walz, R. Ross, L. A. Harker, and J. G. White, Biochemical studies of two patients with the gray platelet syndrome. *J. Clin. Invest. 66*:102-109 (1980).
45. J. Koutts, P. N. Walsh, E. F. Plow, J. W. Fenton, II, B. N. Bouma, and T. S. Zimmerman, Active release of human platelet factor VIII-related antigen by adenosine diphosphate, collagen, and thrombin. *J. Clin. Invest. 62*:1255-1263 (1978).
46. J. W. Slot, B. N. Bouma, R. Montgomery, and T. S. Zimmerman, Platelet factor VIII-related antigen: Immunofluorescent localization. *Thromb. Res. 13*:871-881 (1978).
47. H. Sander, J. W. Slot, B. N. Bouma, and J. Sixma, Immunocytochemical localization of fibrinogen, platelet factor 4 and thromboglobulin in thin frozen sections of human platelets. *J. Clin. Invest. 72*:1277-1287 (1983).
48. J. N. George, R. M. Lyons, and R. K. Morgan, Membrane changes associated with platelet activation. *J. Clin. Invest. 66*:1-9 (1980).
49. J. Lahav and R. O. Hynes, Involvement of fibronectin, von Willebrand factor, and fibrinogen in platelet interaction with solid substrata. *J. Supramol. Struct. Cell Biochem. 17*:299-311 (1981).
50. I. F. Charo, R. D. Feinman, and T. C. Detwiler, Interrelationships of platelet aggregatin and secretion. *J. Clin. Invest. 60*:866-876 (1977).
51. P. M. Henson, Interactin of cells with immune complexes: Adherence, release of constituents and tissue injury. *J. Exp. Med. 134*:1145-1155 (1971).
52. D. C. Morrison, J. F. Roser, C. G. Cochrane, and P. M. Henson, The initiation of mast cell degranulation: Activation at the cell membrane. *J. Immunol. 114*:966-970 (1975).
53. G. O. Gogstad, I. Hagen, R. Korsmo, and N. O. Solum, Evidence for release of soluble, but not of membrane-integrated, proteins from human platelet α-granules. *Biochim. Biophys. Acta 702*:81-89 (1982).
54. G. O. Gostad, I. Hagen, R. Korsmo, and N. O. Solum, Characterization of the proteins of isolated human platelet α-granules. Evidence for a separate granule-pool of the glycoproteins II$_b$ and III$_a$. *Biochim. Biophys. Acta 670*:150-162 (1981).
55. M. Berndt and D. Phillips, *Platelets in Biology and Pathology* (J. Gordon, ed.) (Elseiver, New York, 1980, pp. 43-76.
56. J. Wencel-Drake, T. Kunicki, V. Woods, E. F. Plow, and M. H. Ginsberg, Evidence for distinct intracellular pools of platelet membrane glycoproteins GPIb and GPIIb/III. *Blood (Suppl. 1) 69*:270a (1983).

57. O. Behnke, Electron microscopic observations on the membrane systems of the rat blood platelet. *Anat. Rec. 158*:121-140 (1967).

58. J. G. White, A search for the platelet secretory pathway using electron dense tracers. *Am. J. Pathol. 58*:31-39 (1970).

59. J. C. White and C. Clawson, *Am. J. Pathol. 101*:353 (1980).

60. M. H. Ginsberg, L. Taylor, and R. G. Painter, The mechanism of thrombin-induced platelet factor 4 secretion. *Blood 55*:661-668 (1980).

61. D. Zucker-Franklin, Endocytosis of human platelets: Metabolic and freeze-fracture studies. *J. Cell Biol. 91*:706-715 (1982).

62. R. Painter and M. H. Ginsberg, Centripetal myosin redistribution in thrombin stimulated platelets. Relationship to platelet factor 4 secretion. *Exp. Cell Res. 155*:198-212 (1984).

63. R. Allen, personal communication (1982).

64. J. Flanagan and G. L. E. Koch, Cross-linked surface Ig attaches to actin. *Nature 273*:278-281 (1978).

65. D. R. Phillips, L. K. Jennings, and H. H. Edwards, Identification of membrane proteins mediating the interaction of human platelets. *J. Cell Biol. 86*:77-86 (1980).

66. M. B. Zucker and N. C. Masiello, The triton X-100-insoluble residue ("cytoskeleton") of aggregated platelets contains increased lipid phosphorus as well as ^{125}I-labeled glycoproteins. *Blood 61*:676-683 (1983).

67. R. G. Painter and G. H. Ginsberg, Concanavlin A induces interactions between surface glycoproteins and the platelet cytoskeleton. *J. Cell Biol. 92*:565-573 (1982).

68. A. T. Nurden and J. P. Caen, The different glycoprotein abnormalities in thrombasthenic and Bernard-Soulier platelets. *Semin. Hematol. 16*:234-250 (1979).

69. A. L. Lightsey, W. J. Thomas, E. F. Plow, R. McMillan, and M. Ginsberg, Glanzmann's thrombasthenia in the absence of GPIIb/III deficiency. *Blood (Suppl.) 58*:199a (1981).

70. L. K. Jennings and D. R. Phillips, Purification of glycoproteins IIb and III from human platelet plasma membranes and characterization of a calcium dependent glycoprotein IIb-III complex. *J. Biol. Chem. 257*: 10458-10466 (1982).

71. M. H. Ginsberg, J. Forsyth, A. Lightsey, J. Chediak, and E. F. Plow, Reduced surface expression and binding of fibronectin by thrombin-stimulated thrombasthenic platelets. *J. Clin. Invest. 71*:619-624 (1983a).

72. M. H. Ginsberg, J. Chediak, A. Lightsey, and E. F. Plow, Thrombasthenic platelets do not bind plasma fibronectin. *Thromb. Haemost. 46*:246a (1981).

73. M. B. Donati, G. Balconi, G. Remuzzi, R. Borgia, L. Morasca, and G. de Gaetano, Skin fibroblasts from a patient with Glanzmann's thrombasthenia do not induce fibrin clot retraction. *Thromb. Res. 10*:173-174 (1977).

74. B. M. Steinberg, K. Smith, M. Colozzo, and R. Pollack, Establishment and transformation diminish the ability of fibroblasts to contract a native collagen gel. *J. Cell Biol. 87*:304-308 (1980).
75. J. S. Bennett and G. Vilaire, Exposure of platelet fibrinogen receptors by ADP and epinephrine. *J. Clin. Invest. 64*:1393-1401 (1979).
76. E. F. Plow, G. Marguerie, and M. H. Ginsberg, Fibronectin binding to thrombin-stimulated platelets. *Blood* in press.
77. J. Lahav, M. A. Schwartz, and R. O. Hynes, Analysis of platelet adhesion with a radioactive chemical crosslinking reagent: Interaction of thrombospondin with fibronectin and collagen. *Cell 31*:253-262 (1982).
78. J. Hawiger, S. Parkinson, and S. Timmons, Prostacyclin inhibits mobilisation of fibrinogen-binding sites on human ADP- and thrombin-tested platelets. *Nature 282*:195-197 (1980).
79. E. F. Plow and G. A. Marguerie, Participation of ADP in the binding of fibrinogen to thrombin-stimulated platelets. *Blood 56*:553-555 (1980).
80. Z. M. Ruggeri, R. Bader, and L. De Marco, Glanzmann thrombasthenia: Deficient binding of von Willebrand factor to thrombin-stimulated platelets. *Proc. Natl. Acad. Sci. USA 79*:6038-6041 (1982).
81. M. H. Ginsberg, R. Wolff, B. Coller, R. McEver, G. Marguerie, and E. F. Plow, Thrombospondin binding to thrombin-stimulated platelets: Evidence for a common adhesive protein binding mechanism. *Clin. Res.* in press.
82. R. W. Bunting, E. I. Peerschke, and M. B. Zucker, Human platelet sialic acid content and tritium incorporation after ADP-induced shape change and aggregation. *Blood 52*:643-653 (1978).
83. G. Pietu, G. Cherel, G. Marguerie, and D. Meyer, A new role for fibrinogen: Inhibition of von Willebrand factor-platelet interaction. *Nature* in press.
84. S. Timmons, M. Kloczewiak, and J. Hawiger, Common receptor mechanism for binding of von Willebrand factor and fibrinogen to human platelets. *Blood (Suppl. 1) 62*:269a.
85. E. F. Plow, A. Srouji, D. Meyer, G. Marguerie,and M. H. Ginsberg, Evidence that three adhesive proteins interact with a common recognition site on activated platelets. *J. Biol. Chem. 259*:5388-5391 (1984).

9

Phagocytosis

Livingston VanDeWater III
Beth Israel Hospital
and Harvard Medical School
and the Charles A. Dana Research Institute
Boston, Massachusetts

Phagocytosis is the process by which particles (including fibrin aggregates, tissue debris, dead cells, and bacteria) are bound and ingested by specialized cells, primarily neutrophils and macrophages. The former are blood-borne, migratory cells; the latter cell type is present in the body as both wandering and immobile cells and with monocytes forms the mononuclear phagocyte system [1]. Recognition of particles by phagocytic cells is facilitated by humoral factors (opsonins), which bind to particles and promote their uptake [2]. Immunoglobulins and some complement components are known to have opsonic activity for a wide spectrum of antigenic particles. Recently, an α_2-globulin from plasma has been implicated in the clearance of particles from the circulation, and this protein has been identified as fibronectin [3-5].

Fibronectin is a molecule that interacts with many cell types
and has binding sites for native and denatured collagen, heparin
sulfate, fibrinogen and fibrin, some microorganisms, DNA, actin,
and proteins with collageneous sequences (Clq and acetylcholine-
sterase) and can be cross-linked to fibrin or collagen by factor
XIIIa (reviewed in Refs. 6-9). Fibronectin is found as insoluble
(such as cell surface) or soluble (such as plasma) proteins. The presence of a plasma protein in relatively high concentration (300 µg/
ml in humans) with such an array of binding sites has encouraged
the hypothesis that fibronectin is an opsonin. The work has taken
several directions, including examination of its fluctuation during
disease, its adherence to bacteria and other ligands, and its promotion of particle uptake by defined cells in vitro. The purpose of
this chapter is to review these principal areas of experimentation
and to identify areas for future work.

PLASMA FIBRONECTIN LEVELS AND THE MONONUCLEAR
PHAGOCYTE SYSTEM

Of the mononuclear phagocytes, those present as immobile cells in
organs, including liver and spleen, comprise a system that is largely
responsible for the clearance of particles from the blood. Colloidal
gold or carbon injected intravenously are rapidly removed from
circulation and appear in phagocytic cells lining the organ vasculature (such as Kupffer cells in liver). In hemostasis, Kupffer cells
are an important site for the removal of fibrin aggregates. Lee and
McCluskey [10] showed by immunofluorescence that fibrin was
present in Kupffer cells after injection of thrombin or endotoxin.
Experiments in which rat liver was perfused with a combination of
fibrin and fibrin degradation products also showed that the isolated liver could bind these proteins [11]. To date, no evidence
has been reported showing that purified liver Kupffer cells ingest
fibrin. However, soluble fibrin complexes are bound by peritoneal
macrophages [12-15].

Ingestion of a large colloidal load into an experimental animal
results in a period of phagocytic depression or blockade in which
reduced clearance rates for a second particle challenge are observed [16]. This experimental phagocytic depression has been observed with many clinical conditions, including surgery, trauma,

severe burns, septicemia, and cancer, and is thought to be caused
by circulating tissue debris [5]. Activation of intravascular coagulation is also associated with decreased phagocytic clearance and
has been correlated with excessive fibrin deposition in the renal
and pulmonary vasculature [17, 18]. Thus, phagocytic depression
in patients with intravascular coagulation may aggravate their con-
dition with decreased clearance resulting in organ damage. It is important for improved clinical care to delineate the physiological
elements responsible. Multiple factors may contribute to the observed reduction in clearance rate, including Kupffer cell damage,
impaired liver blood flow, and depletion of available opsonin.

Several lines of evidence show that opsonin depletion is important in the development of phagocytic blockade. Early experiments [19] indicated that crude serum fractions were limiting in
the process of uptake during blockade and that, when supplied intravenously, clearance was partially restored. Furthermore, if test
particles pretreated with a fraction of rat plasma were injected
into animals with depressed phagocytic function, enhanced clearance was observed [20-22].

Recent work has revealed decreased concentration of an α_2-
globulin in patients with surgery trauma, burns, sepsis, or disseminated intravascular coagulation [23-27]. This α_2 protein was
shown to be plasma fibronectin [3, 4]. Depressed fibronectin
levels have also been observed in experimental animals with thrombin-induced intravascular coagulation, Rocky Mountain spotted
fever, major traumatic shock, and burn injury [28-30]. The blood
plasma levels of fibronectin have been measured by immunochemical methods and by bioassay. Immunochemical assay, although
dependent upon the specificity of the antibody used, provides accurate and reproducible quantitation of a specific antigen. With
this type of assay a strong correlation has been observed between
decreased fibronectin levels and phagocytic blockade induced in
experimental animals injected with colloids [31]. This correlates
with the observed decrease of fibronectin in animals or humans
with the diseases just mentioned. However, proof that fibronectin
depletion causes phagocytic blockade would only come in experiments in which plasma fibronectin could be selectively removed
from animals subsequently challenged with a variety (such as gelatin, fibrin, or bacteria) of particles. One would expect prolonged

clearance rates when fibronectin was required for clearance of a particular type of particle. Evidence reported before the identity of fibronectin was known supports this prediction. Delayed clearance of gelatin-colloid was observed when rats were first injected with rabbit anti-α_2-globulin compared with normal rabbit serum or anti-rat albumin [32]. These experiments suggest the importance of fibronectin but should now be expanded with highly defined immunological reagents.

An additional concern at present is that the observed correlation does not prove that fibronectin functions directly as an opsonin. To examine this issue, a bioassay has been devised in which serum or plasma fibronectin promotes the association of gelatin particles to rat liver slices. Serum samples for assay have been taken from several species of animals [22] and humans [25] and analyzed with the rat liver assay. Evidence that all the serum activity is fibronectin has not been presented. Finally, particle binding but not ingestion has been observed [33-35]. However, several reports have appeared showing that, with purified macrophages, plasma fibronectin does promote binding and/or uptake of gelatinized test particles. Isolated Kupffer cells have not yet been demonstrated to ingest fibronectin-coated test particles in vitro.

Despite the strong correlation of decreased plasma fibronectin concentration with decreased rates of particle clearance, conditions that result in blockade do not always coincide with plasma fibronectin depletion. Grossman et al. [36] have demonstrated a brief decrease followed by progressive elevation of plasma fibronectin and fibrinogen levels in animals after injection of endotoxin in rabbits (generalized Shwartzman reaction) [16] or with intraabdominal abscesses in rats. This observation does not diminish the significance of the correlation but suggests that a complicated balance may exist between the temporal relationship of fibronectin levels to blockade by endotoxin or bacterial sepsis. Indeed, blockade may be followed by a period (within hours) of hyperactivity in which increased phagocytosis is observed [16]. Species differences in the delay and extent of hyperactivity have been reported [37]. The relationship of this switch to the regulation of fibronectin synthesis and degradation is unknown and may differ for bacteria and other colloids.

The hypothesis that fibronectin is important to the proper functioning of the mononuclear phagocyte system has been further supported by reconstitution experiments. Cryoprecipitate (containing fibronectin, factor VIII, and fibrinogen) has been infused in septic surgical and trauma patients [25] in an uncontrolled study. Restoration of fibronectin levels was observed by immunological assay and correlated with an improved clinical state. In animal studies [38], either cryoprecipitate or affinity-purified plasma fibronectin injected intravenously into rats prevented phagocytic blockade compared with controls (colloid alone). Clearance rates were measured using iodinated gelatin-lipid droplets and found to return to normal when animals were infused with cryoprecipitate or purified fibronectin. An increase in plasma fibronectin levels was also detected using immunochemical assay.

These results strongly suggest that plasma fibronectin does have an important role in promoting mononuclear phagocyte function but do not prove that fibronectin is an opsonin. Further studies are needed to analyze the types of particles that may be opsonized by fibronectin, to study isolated tissue macrophages for uptake of these particles, and to determine the factors regulating possible fibronectin synthesis and degradation during phagocytic blockade.

FIBRONECTIN BINDS TO BACTERIA

In addition to having binding sites for native and denatured collagens, collageneous proteins, glycosaminoglycans, fibrinogen, and fibrin, fibronectin also adheres to many bacteria. This observation has prompted an intense effort to examine closely the nature of the interaction and the hypothesis that fibronectin may opsonize these bacteria.

Fibronectin has been observed to bind to several types of bacteria. Kuusela [39] first showed that labeled plasma fibronectin bound to *Staphylococcus aureus*. The presence of unlabeled fibronectin or appropriate dilutions of platelet-poor plasma inhibited the binding of the labeled protein to the same extent, suggesting that the binding was specific. Comparable volumes of *Mycobacterium butyricum* bound small amounts of labeled fibronectin. Others have also observed binding (variable percentages) to *S. aureus* [40-45] and have used this species for detailed studies of the binding parameters. However, other types of bacteria have been examined. Conditioned medium from a fibroblastic cell line contain-

ing labeled cellular fibronectin was incubated with gram-negative
(*Escherichia coli* and *Salmonella typhimurium*) or gram-positive
(*S. aureus* and *Bacillus subtilis*) bacteria or yeast (*Saccharomyces
cerevisiae*) and the bound proteins analyzed by SDS gels [44].
Fibronectin was selectively removed from the medium by each
species. Purified cellular fibronectin also bound, suggesting that
accessory proteins present in medium were not necessary. Other
groups have reported the binding of plasma fibronectin to *Strepto-
coccus pyogenes* [46] and to *Staphyloccus epidermidis* [41, 42],
and minimal binding has been reported for *E. coli* and encapsulated *S. aureus* [42]. *Treponema pallidum* has recently been observed to bind plasma fibronectin using radioactive protein, protein immunoblotting, or immunofluorescence [47]. Minimal binding was observed with the avirulent spirochete *T. phagedinis*
(Reiter). The broad range of microorganisms that bind fibronectin
is impressive; however, what role this may have in the pathogenesis
of infection is unclear. The observation that some bacterial types
bind little fibronectin will aid in understanding the physiological
role of this interaction.

The binding of fibronectin to *S. aureus* has been studied in
the greatest detail. Studies using trypsin-degraded fibronectin have
shown that a 27 kd fragment, purified by ion-exchange chromatography, binds to *S. aureus* [40]. The fragment inhibited the binding of iodinated fibronectin as efficiently as unlabeled plasma
fibronectin on a molar basis. This N-terminal region of fibronectin is known to contain basic amino acids, a heparin binding site,
and the factor XIIIa site (s) [Refs. 6-9]. The factor XIIIa site on
the 27 kd peptide was found to be active, and the peptide was covalently bound to *S. aureus* in the presence of thrombin, calcium,
and factor XIIIa [40]. There has not yet been a report that factor
XIIIa and fibronectin function together in plasma to affect the covalent attachment of fibronectin to bacteria. This enzymatic activity is apparently not absolutely necessary for fibronectin binding to several different species of bacteria or to yeast [39, 44, 47].
In some experiments, essentially irreversible binding of fibronectin to *S. aureus* has been observed in the absence of added factor
XIIIa [43, 45]. What role irreversible binding may have in the
functional activity of fibronectin-bacteria complexes remains to be
determined.

Several strains of *S. aureus* have been used to examine the bacterial site that binds fibronectin. In some experiments, fibronectin has been eluted from washed bacterial pellets by ionic detergents [44] or by chaotropic agents [39, 44], suggesting that a portion of the fibronectin is reversibly bound. Recently, experiments have shown that varying degrees of reversible binding are observed, dependent upon the pH of the medium in which bacteria were grown [42, 43]. The effort to obtain a well-defined dissociation constant has been complicated by the state of aggregation of bacteria and the degree of self-association of fibronectin. An approximate dissociation constant has been reported to be in the nanomolar range with 1,-10,000 sites per cell (depending on growth conditions) [43],which would imply that, in human plasma (fibronectin, 300 μg/ml; 7.5×10^{-7} M), all bacterial sites are occupied. Fibronectin binding appears to be saturable and time dependent and occurs on live or heat-killed bacteria [43, 45]. The degree of binding has been found to be dependent on the growth state and medium composition [43]. Several investigators have observed specificity in binding: fibrinogen, fibronectin-depleted plasma, protein A, collagen fragments (α_1 CB7), dextran, cationic proteins, α_1-fetoprotein do not inhibit binding of labeled fibronectin to bacteria. The *S. aureus* fibronectin receptor has been reported to be removed by trypsin [42, 45] and is absent on mutant cells that have protein A but do not bind fibronectin. It has been isolated by affinity chromatography on fibronectin-agarose and found to be an 19 kd protein [45].

Questions arise about the function of this fibronectin binding site. What is the molecular relationship of this *S. aureus* protein to receptors on other microorganisms? Do all species bind the same 27 kd portion of fibronectin? Is the factor XIIIa site within this *S. aureus* protein? Also, what is the role of factor XIIIa transamidation, and what is the effect of inhibition of fibronectin binding or transamidation in vivo? Does fibronectin affect either positively or negatively the recognition of bacteria as foreign, the clearance by neutrophils or by the mononuclear phagocyte system, or the adherence of bacteria to host tissues (see Refs. 43 and 45)?

Recent evidence has appeared that raises the possibility that fibronectin is an important mediator of reticuloendothelial system function. Depletion of plasma fibronectin has been observed in animals that have undergone surgery, blunt trauma, hemorrhagic

shock, or burn injury, conditions that reduce reticulendothelial
system phagocytic activity. Indeed, low levels of fibronectin, associated with disseminated intravascular coagulation in humans, correlated with a poor prognosis [48]. Fibronectin may facilitate the
clearance on nonbacterial debris arising from trauma or septicemia,
or it could function directly as an opsonin for bacteria by neutrophils or mononuclear phagocytes or indirectly to promote mononuclear phagocyte system function. Recently, patients presenting
with acute leukemia were reported to have low levels of fibronectin associated not with tumor growth or remission but with periods
of bacterial infection [49]. The plasma concentration of fibronectin was found to be lowest in severe infections and returned to
normal with cessation of infection. One case of hyposplenism has
also been reported in which apparent splenic blockade, as detected
by [99m Tc] albumin scan, was correlated with low levels of fibronectin and frank pneumonia. Reversal of the pneumonia by anti-
biotic treatment was accompanied by normal spleen scan and
plasma fibronectin levels [26]. However, experiments with rats
challenged with live *E. coli* showed no similar decrease of plasma
fibronectin concentration [50]. In another study, rats with intraabdominal abscesses or rabbits undergoing the classic generalized
Shwartzmann reaction exhibited elevated concentrations of both
fibrinogen and fibronectin [36]. In the presence of discrepant observations between human and animal studies, it is not possible
at present to state what role fibronectin may have in bacterial sepsis. It may be an opsonin for bacteria, for tissue debris arising from
sepsis, or as some other mediator of mononuclear phagocyte system function or integrity.

    If fibronectin is to be termed an opsonin for bacteria, then it
must be possible to show in vitro that defined phagocytic cells
ingest bacteria coated with fibronectin. Several laboratories have
utilized neutrophils [42, 44, 51, 52] or monocytes [42], alveolar
macrophages [42, 44], or macrophage cell lines [44] as sources
of phagocytic cells. These cells have been incubated with *S. aureus*
[42, 44, 51, 52], *E. coli* [44, 52], or *S. typhimurium* [44]. Lanser and Saba [51] have reported that fibronectin does not specifically promote binding of *S. aureus* to human neutrophils but that
complete serum is slightly more effective as an opsonin than is
fibronectin-depleted serum. They suggest that fibronectin may act

as a cofactor that enhances phagocytosis of *S. aureus* in this system. Simpson et al. [46] observed stimulation of binding of M⁻
but not M⁺ *S. pyogenes* to human neutrophils. Possible phagocytosis was not examined. Verbrugh et al. [42], using an assay for
direct uptake of radiolabeled bacteria by human monocytes, neutrophils, or alveolar macrophages, found that fibronectin depletion
or reconstitution of serum had no effect on bacterial phagocytosis
but depletion of C3 did. Using bactericidal and chemiluminescence
assays, Proctor et al. [52] observed that fibronectin promoted
neutrophil chemiluminescence and adhesion of *S. aureus*, but not
E. coli. Bacteria coated with purified fibronectin were susceptible
to enzymatic digestion (lysostaphin), but serum-coated bacteria
showed progressive resistance to digestion, proving that fibronectin promotes their attachment but is not sufficient for phagocytosis.

In another study, five different phagocytic cell types (alveolar
macrophages, neutrophils, or mouse phagocytes) were mixed in
suspension with labeled *S. aureus* or *S. typhimurium* [44]. It was
observed that both complete and fibronectin-depleted serum promoted uptake of bacteria but heat-inactivated serum did not. Electron microscopy confirmed that bacteria were ingested in the first
case but not in the second. Several variables were examined for
possible influence on the activity of fibronectin. Purification of
plasma or cellular fibronectin by several methods in the presence
or absence of reducing agent did not increase activity. Incubation
of bacterial-phagocyte mixtures with heparin did not stimulate uptake. Depletion of fibronectin from plasma or serum did not significantly reduce the level of bacterial phagocytosis observed, suggesting that fibronectin does not enhance the uptake by some
other opsonin. Macrophage synthesis could not explain the lack of
dependence on added fibronectin because no uptake was observed
in serum-free medium and some of the cells tested make little or
no fibronectin [53]. Despite the observed lack of fibronectin-dependent bacterial uptake, alveolar and peritoneal macrophages,
neutrophils, and macrophagelike cell lines do ingest fibronectin-
coated gelatin latex [53].

Fibronectin has also been suggested to mediate the attachment
of bacteria to host tissues. This has recently been examined in the
case of *T. pallidum*, a spirochete that adheres to host tissue receptors by a tip structure mediated by host plasma proteins and speci-

fic treponemal proteins [47]. *Treponema pallidum* was found to adhere to fibronectin-coated glass or to cells of a human epithelial line and to show decreased binding in the presence of anti-fibronectin. Three treponemal envelope proteins have been found to bind to fibronectin affinity columns. *T. phagedenis* biotype Reiter, an avirulent spirochete, was observed not to bind to fibronectin, strengthening the hypothesized role of fibronectin in bacterial adherence. If fibronectin is involved in the colonization of host tissues by a spectrum of bacteria, then plasma fibronectin may serve to block, by competition, this interaction by its presence in relatively high concentration in the blood [52]. Studies of bacterial binding to defined cells and tissue matrices will contribute to resolving the role of fibronectin in the parasitism of the host by microorganisms.

In summary, there is no direct evidence showing that fibronectin is an opsonin for bacteria. It is possible that other phagocytic cells (splenic or other tissue macrophages) or that more activated forms of the cells tested will be found that do ingest bacteria. Alternatively, fibronectin may facilitate bacterial adhesion to phagocytic cells, resulting in more efficient killing by other mechanisms. Finally, early evidence suggests that some serum factors promote or co-opsonize [54]. If fibronectin is such a factor or if it activates macrophages, the reticuloendothelial system might function less efficiently when fibronectin levels are decreased, thus exhibiting sustained blockade.

PROMOTION OF PARTICLE UPTAKE

A fundamental aspect of the investigation of opsonic activity is the direct demonstration of uptake by defined particles and phagocytic cells. Several test particles have been used, including erythrocytes, yeast cell wall (zymosan), and plastic or latex particles [2]. Each offers distinct advantages, such as morphological appearance, ease of scoring uptake, variations in surface coat, or particle density. Ideally, one hopes that the particle of choice will be advantageous experimentally but also closely resemble a physiological entity. Similar considerations apply to the choice of phagocytic cell. Ideally, a phagocytic cell should be derived from a relevant tissue (such as liver or spleen in the case of sessile macrophages), be homogeneous in phenotype, and be available in abundance. Cloned macrophagelike cell lines provide large numbers of homogeneous cells that can be easily harvested, metabolically labeled, and have some differentiated properties [55]. Because the cells are transformed, however, they cannot be considered identical with macrophages in situ. In practice, a combined approach using a variety of cell types best approximates conditions in vivo.

The hypothesis that fibronectin is an opsonin has been advanced by the use of gelatin coupled to latex particles. These particles are especially useful because the surface can be easily modified to conjugate any protein at a desired surface density to a particle of defined size and density. The initial work on fibronectin was carried out using gelatin-coated oil droplets in combination with rat liver slices [22]. Saba and coworkers developed this bioassay as a method to quantitate the amount of α_2-opsonic protein that promoted the association of gelatin oil droplets to liver cells (presumed to be Kupffer cells), and to show that the active protein was antigenically identical to plasma fibronectin [3, 4]. These results were extended by studies using latex particles coated with gelatin, fibrinogen, fibrin, or serum albumin [34]. A specific association of gelatin-latex particles (not other particles) with liver was observed that was also dependent upon heparin. However, this association occurred without internalization: Trypsin treatment of liver slice with bound particles reduced the level of binding to background, showing that all particles were outside cells. Nor was it directly demonstrated by electron microscopy that gelatin particles are ingested by cells within the liver slice.

Several experiments have been reported in which isolated macrophages have bound gelatin-coated particles. Early work with guinea pig peritoneal exudate cells showed that a high-molecular-weight protein in normal serum promoted the binding of denatured collagen to cells [56]. Doran et al. [35] observed an association of gelatin-latex particles with peritoneal macrophages in the presence of serum or plasma fibronectin that could be blocked by metabolic inhibitors; no direct proof of ingestion was given. Electron microscopy was used by Gudewicz et al. [57] to demonstrate that gelatin-latex particles were ingested by monolayer cultures of rat peritoneal macrophages in the presence of fibronectin. Uptake (over 60 min) was stimulated by, but not absolutely dependent upon, heparin, which was ineffective if fibronectin was omitted. Ingestion was further substantiated by the use of trypsin to remove bound particles. Further characterization of fibronectin-dependent phagocytosis of gelatin-latex has been presented using a macrophage cell line (P388D$_1$) [53]. Cells in suspension or monolayers were incubated with labeled gelatin-latex and were dependent upon fibronectin for uptake; serum depleted of fibronectin did not support bead uptake until reconstituted with purified protein. This uptake was strictly dependent upon fibronectin concentration, gelatin density on the latex, and heparin concentration. Cellular fibronectin was also observed to promote uptake (VanDeWater and Hynes, unpublished) in agreement with Marquette et al. [58]. Ingestion was proven by electron microscopy of sectioned cells from the suspension assays and by an immunofluorescent technique for monolayers. Furthermore, the nature of the dependence on heparin is unknown; among the possibilities are the necessity of forming more platable aggregates between fibronectin, heparin and gelatin, which may recruit greater numbers of macrophage receptors and/or induce a conformation change in fibronectin [33, 59–62]. The observed dependence on fibronectin was absolute because this cell line does not synthesize fibronectin [53].

Human alveolar macrophages have also been analyzed for their ability to ingest gelatin-latex [63]. These cells are interesting because they synthesize and secrete large amounts of fibronectin and may be largely dependent upon this fibronectin for phagocytosis in vivo [64, 65]. However, added fibronectin was needed in vitro to direct phagocytosis of gelatin beads by washed, macrophage monolayers in a concentration-dependent fashion. Resistance of particles to trypsinization after incubation of cells proved that beads were phagocytosed by the macrophages. Microscopy of sectioned cell monolayers confirmed that particles were ingested. Similar experiments using albumin-latex indicated that this was specific for gelatin. Domain-specific antibodies were used to show that both the gelatin and the cell-binding regions of fibronectin were required for uptake. Inhibition studies with antibodies provide important proof that fibronectin, not a contaminant, is the opsonin. Heparin was not required for gelatin-latex uptake with this cell type. In summary, macrophages from different sources do ingest gelatin particles, providing a framework for closer analysis of this opsonic activity in vivo. Fibronectin has also been shown to enhance the uptake of zymosan and gelatin-latex by human neutrophils [66].

Several points merit further comment. First, the reason for variable dependence upon added heparin is unknown. The difference in the degree of dependence upon heparin could be due to variable production of polyanionic species by different macrophages, either secreted or at the cell surface. Alternatively, macrophage surfaces could be differentially responsive to heparin-fibronectin complexes. The data of McDonald et al. [63] suggest that, with human alveolar macrophages, the uptake of gelatin-latex particles is mediated by a fibronectin-cell interaction, not a heparin-cell interaction. A trivial explanation for the observed dependence might be aggregation of particles, thereby increasing the number bound. However, with P388D$_1$ [53], no uptake occurred without heparin so that its role cannot only be an apparent enhancement of particle binding by simply increasing the size of the bound aggregate [33, 59]. The role of the heparin region will be determined in part by the use of domain-specific antibodies. If one blocks this region with antibodies either before or after aggregate formation, what is the effect on uptake? In any event, resolution of this issue will aid in the understanding of the importance of glycosaminoglycans on the role of fibronectin as an adhesive protein.

Second, the relationship of fibronectin-dependent uptake to nonspecific phagocytosis of plain latex [2] is unclear. On the one hand, nonspecific ingestion of latex usually takes place in growth medium supplemented with serum so that uptake could be mediated by any serum proteins that adhere to plastic (such as IgG, C3, and fibronectin) and bind to the macrophage. These proteins adhere with different affinity and different density so that their uptake by macrophages could be viewed as either nonspecific or multispecific. On the other hand, carboxylate-modified latex to which gelatin or other proteins are covalently bound produces a different particle, with a defined composition. The interaction of fibronectin with these gelatin particles is specific. Indeed, one group showed a strong dependence on the density of gelatin on latex to get efficient fibronectin-mediated uptake of particles [53]. A related by separated issue concerns the specificity of fibronectin as an opsonin for all protein-coated particles. A recent report shows that fibronectin promotes the uptake of gelatin-latex, but not albumin-latex [63]. At present, the data suggest that fibronectin functions selectively for certain protein-coated particles.

Third, serum-coated zymosan has been found to enhance macrophage chemiluminescence and glucose oxidation to a much greater extent than fibronectin-coated zymosan or gelatin-latex [66]. A mechanism may exist for discriminating the type of particle as well as the opsonin. A recent investigation with the protozoan amoeba *Acanthaamoeba castellini* has shown that this phagocyte avidly ingests latex or zymosan particles but discriminates between digestible or indigestible particles [67]. Furthermore, recent evidence shows that different intracellular messages mediate whether phagocytosis is followed by peroxide release. Human monocytes liberate peroxide when IgG-coated particles, but not C3b- or C3bi-coated particles, are ingested [68]. Monocytes attached to fibronectin-coated dishes have enhanced uptake of particles opsonized with C3b. It is important to understand how these selective mechanisms function and what role fibronectin may have in modulating phagocytosis.

Reports have appeared using human monocytes that suggest that fibronectin is not active as an opsonin for gelatin-coated particles with these cells. Bevilacqua et al. [69] found that monocytes could bind but were unable to ingest fibronectin-coated gelatin-latex or gelatin-erythrocytes. Monocytes were observed to attach to fibronectin-coated gelatin-treated dishes and under these conditions exhibited enhanced attachment and phagocytosis of IgG-coated particles and enhanced attachment of C3b-particles. Related results have been obtained using human monocytes in suspension or monolayer cultures [70]. Enhanced attachment and phagocytosis was observed with IgG- or C3b-coated particles, on cells in suspension or monolayer cultures.

However, others have shown that, with human monocytes, the uptake of C3b-particles is stimulated to a lesser degree by soluble than with surface-bound fibronectin [71]. Interestingly, the observed enhancement in C3b-particle uptake on the apical surface area occurs with fibronectin beneath the cells, suggesting a functional but not necessarily a physical linkage between receptors in the cells.

The results with monocytes contrast with those using macrophages, in which direct uptake of fibronectin-coated particles is

observed [53, 57-59, 63, 66]. The nature of this discrepancy is unclear. Perhaps monocytes do not have sufficient numbers or density of fibronectin-binding sites to mediate vigorous phagocytosis
of fibronectin-particles. Wright et al. [71] observed that, with prolonged culture of human monocytes, the enhancement by fibronectin of the phagocytosis of C3b-particles was increased with
time in culture, suggesting that the process might be developmen-
tally regulated. The exact relationship between monocytes in long-
term culture and elicited macrophages is unknown. Perhaps developing monocytes undergo a series of steps toward full activation
that enable phagocytosis of IgG- or C3b-particles early but phagocytosis of fibronectin-particles later. These latter stages might include an increase in the number of fibronectin binding sites or
changes in their structure or connection with the cytoskeleton.

   Aside from the importance of determining the extent of fibronectin's role as an opsonin, the endocytosis of fibronectin-
coated particles by cells in culture also provides a means for studying fibronectin-membrane interactions. Particle-containing vesicles
can be isolated from phagocytic cells and analyzed for cell surface
proteins [72]. These regions or domains of the membrane may
differ for various opsonins, and it is important to compare them
within the same macrophage cell type or between cell types (fibroblasts and macrophages [73-76] in the case of fibronectin). In
fibroblasts, an association has been demonstrated between extracellular fibronectin and intracellular cytoskeletal proteins [73]. If
the transmembrane connection between fibronectin and actin is
retained by isolated phagosomes, it should be possible to isolate
phagosomes from fibroblasts or macrophages and analyze this
transmembrane link [77].

   Studies using defined particles, such as gelatin-latex, have
shown that isolated macrophages phagocytose particles opsonized
with fibronectin. This ingestion is dependent upon fibronectin
concentration and variably dependent upon heparin and has been
observed with elicited macrophages but not with monocytes. To
better understand the importance of these observations of pathological processes, such as thrombosis, inflammation, and sepsis,
relevant proteins that also bind to fibronectin (such as fibrin, with
or without factor XIIIa, collagens, or Clq) [78] should be examined
for uptake by macrophages.

SUMMARY

There has been great interest in the possibility that plasma fibronectin is an opsonin for bacteria, or for the products of throm-
botic or inflammatory processes. Evidence both for and against
this hypothesis has been presented in this chapter. On the one
hand, isolated macrophages from different species have been
shown to ingest gelatin-coated particles in a fibronectin-dependent
fashion. Indirect evidence suggests that particles known to interact
with fibronectin (such as gelatin emulsions or fibrin) are found to
be ingested by hepatic or splenic macrophages in vivo. Furthermore, fibronectin levels are depressed in experimental models in
which large doses of gelatinized test colloid are injected and decreased particle clearance rates are observed. Finally, depressed
phagocytic function in test animals can be restored by the infusion of cryoprecipitate or plasma fibronectin. An improved clinical
state has also been observed in some patients. On the other hand,
fibronectin binds to such particles as bacteria, but it does not promote their ingestion, showing that the simple binding of the protein is not sufficient for uptake by the neutrophils or macrophages
tested. Furthermore, although macrophages and neutrophils do ingest gelatin-coated particles, monocytes apparently do not. Finally,
although depressed fibronectin levels are often associated with
phagocytic depression, in some animal models (such as sepsis and
Shwartzman reaction) prolonged decreases in fibronectin levels are
not observed.

Consequently, it is premature to conclude that fibronectin is
an opsonin similar to IgG or C3b. To understand fully the specificity of the interaction between particle-bound fibronectin and
phagocytic cells, the binding site for fibronectin must be characterized. Is this site required for particle uptake, how does it interact with the cytoskeleton to promote endocytosis, and is it different in composition or quantity on monocytes compared to
macrophages? Also, molecules known to interact with fibronectin,
which are also important products of hemostasis, inflammation, or
wound repair, must be examined for uptake by a broader range of
isolated and mobile and sessile macrophages. Perhaps differences
in uptake will be discerned among populations of cells. With an increased understanding of the fine structure of fibronectin and its
cellular receptor it should be possible to ultilize specific antibody

fragments to determine if selective inhibition of clearance occurs for relevant stromal proteins.

ACKNOWLEDGMENTS

I would like to thank Norma Casner and Sherry Rolley for typing the manuscript and Drs. Harold Dvorak, Richard Hynes, Martin Schwartz, and Donald Senger for their constructive criticism.

REFERENCES

1. R. Van Furth, Cells of the mononuclear phagocyte system, nomenclature in terms of sites and conditions, in *Mononuclear Phagocytes: Functional Aspects* (R. Van Furth, ed.). Martinus Nijhoff, The Hague, 1980, pp. 1-30.
2. S. C. Silverstein, R. M. Steinman, and Z. A. Cohn, Endocytosis, *Ann. Rev. Biochem. 46*:669-722 (1977).
3. F. Blumenstock, P. Weber, and T. M. Saba, Isolation and biochemical characterization of α_2 opsonic glycoprotein from rat serum. *J. Biol. Chem. 252*:7156-7162 (1977).
4. F. Blumenstock, T. M. Saba, P. Weber, and R. Laffin, Biochemical and immunological characterization of human opsonic α_2 SB glycoprotein: Its identity with cold-insoluble globulin, *J. Biol. Chem. 253*:4287-4291 (1978).
5. T. M. Saba and E. Jaffe, Plasma fibronectin (opsonic glycoprotein): Its synthesis by vascular endothelial cells and role in cardiopulmonary integrity after trauma as related to reticuloendothelial function, *Am. J. Med. 68*:577-594 (1980).
6. R. O. Hynes, Fibronectin and its relation to cell structure and behavior, in *Cell Biology of the Extracellular Matrix* (E. D. Hay, ed.). Plenum, New York, 1981, pp. 295-333.
7. D. F. Mosher, Fibronectin, *Prog. Hemost. Thromb. 5*:111-151 (1980).
8. E. Ruoslahti, E. Engvall, and E. G. Hayman, Fibronectin current concepts of its structure and function, *Collagen Res. 1*:95-128 (1981).
9. K. M. Yamada, Biochemistry of fibronectin, in *The Glycoconjugates*, Vol. III, (M. I. Horowitz, ed.). Academic Press, New York, 1982.
10. L. Lee and R. T. McCluskey, Immunohistochemical demonstration of the reticuloendothelial clearance of circulating fibrin aggregates, *J. Exp. Med. 116*:611-613 (1962).
11. H. Gans and J. Lowman, Uptake of fibrin and fibrin degradation products by the isolated perfused rat liver. *Blood 29*:526-539 (1967).
12. L. A. Sherman and J. Lee, Specific binding of soluble fibrin to macrophages. *J. Exp. Med. 145*:76-85 (1977).

13. R. B. Colvin and H. F. Dvorak, Fibrinogen/fibrin on the surface of
    macrophages: Detection, distribution, binding requirements, and pos-
    sible role in macrophage adherence phenomena, *J. Exp. Med.* 142:1377-
    1390 (1970).
14. F. Jilek and H. Hörmann, Fibronectin. V. Mediation of fibrin monomer
    binding to macrophages, *Hoppe-Seylers Z. Physiol. Chem.* 359:1603-
    1605 (1978).
15. S. R. Gonda and J. R. Shainoff, Adsorptive endocytosis of fibrin mono-
    mer by macrophages: Evidence of a receptor for the amino terminus of
    the fibrin α chain, *Proc. Natl. Acad. Sci. USA* 79:4565-4569 (1982).
16. B. Benacerraf, G. Biozzi, B. N. Halpern, and C. Stiffel, Physiology of
    phagocytosis of particles by the RES, in *Physiology of Reticulendothelial
    System* (B. N. Halpern, ed.). Charles C. Thomas, Springfield, Illinois,
    1967, pp. 52-79.
17. L. Lee, Reticuloendothelial clearance of circulating fibrin in the patho-
    genesis of the generalized Schwartzman reaction, *J. Exp. Med. 115:*
    1065-1082 (1962).
18. J. E. Kaplan, The role of the reticuloendothelial system in control of
    hemostatic and thrombotic mechanisms, in *Pathophysiology of the
    Reticuloendothelial System* (B. Altura and T. M. Saba, eds.). Raven
    Press, New York, 1981, pp. 111-129
19. C. R. Jenkin and D. Rowley, The role of opsonins in the clearance of
    living and inert particles by cells of the reticuloendothelial society. *J.
    Exp. Med.* 114:363-371 (1961).
20. I. M. Murray, The mechanism of blockade of the reticuloendothelial
    system. *J. Exp. Med.* 117:139-147 (1963).
21. S. J. Normann and E. P. Benditt, Function of the reticuloendothelial
    system. II. Participation of a serum factor in carbon clearance, *J. Exp.
    Med.* 122:709-719 (1965).
22. T. M. Saba and DiLuzio, Reticuloendothelial blockade and recovery as a
    function on opsonic activity, *Am. J. Physiol.* 216:197-205 (1969).
23. T. M. Saba, W. A. Scovill, and S. P. Power, Human host defense mechan-
    isms as they relate to surgery and trauma. *Surg. Annu.* 12:1-20 (1975).
24. W. A. Scovill, T. M. Saba, F. Blumenstock, H. Bernard, and S. R. Powers,
    Opsonic α₂ surface binding glycoprotein therapy during sepsis, *Ann.
    Surg.* 188:521-529 (1978).
25. T. M. Saba, F. A. Blumenstock, W. A. Scovill, and H. Bernard, Cryopre-
    cipitate reversal of opsonic α₂-surface binding glycoprotein deficiency in
    septic surgical and trauma patients, *Science* 201:622-624 (1978).
26. B. J. Boughton, A. Simpson, and S. Chandler, Functional hyposplenism
    during pneumococcal septicemia, *Lancet i:*121-122 (1983).
27. D. F. Mosher and E. M. Williams, Fibronectin concentration is decreased
    in plasma of severely ill patients with disseminated intravascular coagula-
    tion, *J. Lab. Clin. Med.* 91:729-735 (1978).

28. J. E. Kaplan and T. M. Saba, Low-grade intravascular coagulation and re-ticuloendothelial function, *Am. J. Physiol. 234*:H323-329 (1978).
29. D. F. Mosher, Changes in plasma cold insoluble globulin concentration during experimental Rocky Mountain Spotted Fever infection in Rhesus monkeys, *Thromb. Res. 9*:37-45 (1976).
30. Kaplan and T. M. Saba, Humoral deficiency and reticuloendothelial depression after traumatic shock. *Am. J. Physiol. 230*:7-14 (1976).
31. F. Blumenstock, P. Weber, T. M. Saba, and R. Laffin, Electroimmunoassay of alpha-2-opsonic protein during reticuloendothelial blockade. *Am. J. Physiol. 232*:R80-R87 (1977).
32. J. E. Kaplan, T. M. Saba, and E. Cho, Serological modifications of reticuloendothelial capacity and altered resistance to traumatic, *Circ. Shock 3*:203-206 (1976).
33. I. J. Check, H. C. Wolfman, T. B. Coley, and R. L. Hunter, Agglutination assay for human opsonic factor using gelatin-coated latex particles, *J. Reticuloendothel. Soc. 25*:351-362 (1979).
34. J. Molnar, F. B. Gelder, M. Z. Lai, G. E. Siefrig, Jr., R. B. Credo, and L. Lorand, Purification of opsonically active human and rat cold-insoluble globulin (plasma fibronectin), *Biochemistry 18*:3909-3916 (1979).
35. J. E. Doran, A. R. Mansbogen, and A. C. Reese, Cold-insoluble globulin-enhanced phagocytosis of gelatinized targets by macrophage monolayers: A model system, *J. Reticuloendothel. Soc. 27*:471-483 (1980).
36. J. Grossman, T. Pohlman, F. Koerner, and D. Mosher, Plasma fibronectin concentration in animal models of sepsis and endotoxemia, *J. Surg. Res. 34*:145-150 (1983).
37. G. D. Niehaus, P. R. Schumacker, and T. M. Saba, Reticuloendothelial clearance of blood-borne particulates, *Ann. Surg. 191*:479-487 (1980).
38. T. M. Saba and E. Cho, Reticuloendothelial systemic response to operative trauma as influenced by cryoprecipitate or cold-insoluble globulin therapy, *J. Reticuloendothel. Soc. 26*:171-186 (1979).
39. P. Kuusela, Fibronectin binds to *Staphylococcus aureus. Nature 276*:718-720 (1978).
40. D. F. Mosher and R. A. Proctor, Binding and Factor XIIIa-mediated cross-linking of a 27 kilodalton fragment of fibronectin to *Staphylococcus aureus. Science 209*:927-929 (1980).
41. J. E. Doran and R. H. Raynor, Fibronectin binding to protein A-containing staphylococci. *Infect. Immun. 33*:683-689 (1981).
42. H. A. Verbrugh, P. K. Peterson, D. F. Smith, B.-Y. T. Nguyen, J. R. Hoidal, B. J. Wilkinson, J. Verhoef, and L. T. Furcht, Human fibronectin binding to Staphylococcal surface protein and its relative inefficiency in promoting phagocytosis by human polymorphonuclear leukocytes, monocytes and alveolar macrophages. *Infect. Immun 33*:811-819 (1981).

43. R. A. Proctor, D. F. Mosher, and P. J. Olbrantz, Fibronectin binding to *Staphylococcus aureus. J. Biol. Chem. 257*:14788-14794 (1982).
44. L. VanDeWater, A. T. Destree, and R. O. Hynes, Fibronectin binds to some bacteria but does not promote their uptake by phagocytic cells. *Science 220*:201-204 (1983).
45. C. Ryden, K. Rubin, P. Speziale, M. Höök, M. Lindberg, and T. Wadstrom, Fibronectin receptors from *Staphylococcus aureus. J. Biol. Chem. 258*:3396-3401 (1983).
46. W. A. Simpson, D. L. Hasty, J. M. Mason, and E. H. Beachey, Fibronectin-mediated binding of group A Streptococci to human polymorphonuclear leukocytes. *Infect. Immun. 37*:805-810 (1982).
47. K. M. Peterson, J. B. Baseman, and J. F. Alderete, *Treponema pallidum* receptor binding protein interact with fibronectin. *J. Exp. Med. 157*: 1958-1970 (1983).
48. D. F. Mosher and E. M. Williams, Fibronectin concentration is decreased in plasma of severely ill patients with disseminated intravascular coagulation. *J. Lab. Clin. Med. 91*:729-735 (1978).
49. B. J. Boughton and A. Simpson, Plasma fibronectin in acute leukemia. *Br. J. Hematol. 51*:487-491 (1982).
50. J. E. Kaplan, W. A. Scovill, and H. Bernard, Reticuloendothelial phagocytic response to bacterial challenge after traumatic shock. *Circ. Shock 4*:1-10 (1977).
51. M. F. Lanser and T. M. Saba, Fibronectin as a co-factor necessary for optimal granulocyte phagocytosis of *Staphylococcus aureus. J. Reticuloendothel. Soc. 30*:415-424 (1981).
52. R. A. Proctor, E. Prendergast, and D. F. Mosher, Fibronectin mediates attachment of *Staphylococcus aureus* to human neutrophils. *Blood 59*: 681-687 (1982).
53. L. VanDeWater, III, S. Schroeder, E. B. Crenshaw, III, and R. O. Hynes, Phagocytosis of gelatin-latex particles by a murine macrophage line is dependent on fibronectin and heparin. *J. Cell Biol. 90*:32-39 (1981).
54. G. H. Stollerman, M. Rytel, and J. Ortiz Accessory plasma factors involved in the bactericidal test for type-specific antibody to group A Streptococci. II. Human plasma cofactor(s) enhancing opsonization of encapsulated organisms. *J. Exp. Med.* 1-17 (1962).
55. P. Ralph, Functions of macrophage cell lines, in *Mononuclear Phagocytes: Functional Aspects* (R. Van Furth, ed.). Martinus Nijhoff, The Hague, 1980, pp. 439-456.
56. K. E. Hopper, B. C. Adelman, C. Gentner, and S. Gay, Recognition by guinea pig peritoneal exudate cells of conformationally different states of the collagen molecule. *Immunology 30*:249-259 (1976).
57. P. W. Gudewicz, J. Molnar, M. Z. Lai, D. W. Beezhold, G. E. Siefring, Jr., R. B. Credo, and L. Lorand, Fibronectin-mediated uptake of gelatin-coated latex particles by peritoneal macrophages. *J. Cell. Biol. 87*:427-433 (1980).

PHAGOCYTOSIS 195

58. D. Marguette, J. Molnar, K. Yamada, D. Schlesinger, S. Darby, and P. Van Allen, Phagocytosis-promoting activity of avian plasma and fibroblastic cell surface fibronectins. *Mol. Cell. Biochem.* 36:147-155 (1981).
59. J. E. Doran, A. R. Mansberger, H. T. Edmondson, and A. C. Reese, Cold-insoluble globulin and heparin interactions in phagocytosis by macrophage monolayers: Mechanism of heparin enhancement. *J. Reticuloendothel. Soc.* 29:285-294 (1981).
60. S. Johansson and M. Höök, Heparin enhances the rate of binding of fibronectin to collagen. *Biochem. J.* 187:521-524 (1980).
61. F. Jilek and H. Horman, Fibronectin (cold-insoluble) globulin. VI. Influence of heparin and hyaluronic acid on the binding of native collagen, *Hoppe-Seylers Z. Physiol. Chem.* 360:597-603 (1979).
62. N. E. Stathakis and M. W. Mosesson, Interactions among heparin cold-insoluble globulin, and fibrinogen in formation of the heparin-precipitable fractions of plasma. *J. Clin. Invest.* 60:855-865 (1977).
63. J. A. McDonald, B. Villiger, D. G. Kelley, and T. J. Broekelmann, Analysis of fibronectin mediated gelatin-latex particle binding and uptake by human alveolar macrophages using purified fibronectin domains and domain specific antibodies as inhibitors, submitted for publication (1984).
64. B. Villiger, T. Broekelmann, D. Kelley, G. J. Heymach, III, and J. A. McDonald, Bronchoalveolar fibronectin in smokers and nonsmokers, *Am. Rev. Respir. Dis.* 124:652-654 (1981).
65. B. Villiger, D. G. Kelley, W. Engleman, C. Kuhn, III, and J. A. McDonald, Human alveolar macrophage fibronectin: Synthesis, secretion and ultrastructural localization during gelatin-coated latex particle binding. *J. Cell. Biol.* 90:711-720 (1981).
66. P. W. Gudewicz, D. H. Beezhold, P. Van Alten, and J. Molnar, Lack of stimulation of post-phagocytic metabolic activities of polymorphonuclear leukocytes by fibronectin opsonized particles. *J. Reticuloendothel. Soc.* 32:143-154 (1982).
67. B. Bowers and T. E. Olszewski, *Acanthamoeba* discriminates internally between digestible and indigestible particles. *J. Cell Biol.* 97:317-322 (1983).
68. S. D. Wright and S. C. Silverstein, Receptors for C3b and C3bi promote phagocytosis but not the release of toxic oxygen from human phagocytes. *J. Exp. Med.* 158:2016-2023 (1983).
69. M. P. Bevilacqua, D. Amrani, M. W. Mosesson, and C. Bianco, Receptors for cold-insoluble globulin (plasma fibronectin) on human monocytes. *J. Exp. Med.* 153:42-60 (1981).
70. C. G. Pommier, S. Inada, L. F. Fries, T. Takahashi, M. M. Frank, and E. J. Brown, Plasma fibronectin enhances phagocytosis of opsonized particles by human peripheral blood monocytes. *J. Exp. Med.* 157:1844-1845 (1983).

71. S. D. Wright, L. S. Craigmyle, and S. C. Silverstein, Fibronectin and serum amyloid P component stimulate C3b- and C3bi-mediated phagocytosis in cultured human monocytes, *J. Exp. Med. 158*:1338-1343 (1983).

72. M. G. Wetzel and E. D. Korn, Phagocytosis of latex beads by *Acanthamoeba* castellanii (Neff). III. Isolation of the phagocytic vesicles and their membranes. *J. Cell Biol. 43*:90-97 (1969).

73. R. O. Hynes, A. T. Destree, and D. D. Wagner, Relationships between microfilaments, cell-substratum adhesion and fibronectin. *Cold Spring Harbor Symp. Quant. Biol. 46*:659-670 (1982).

74. F. Grinnell, Fibroblast receptor for cell-substratum adhesion: Studies on the interaction of baby hamster kidney cells with latex beads coated by cold insoluble globulin (plasma fibronectin). *J. Cell Biol. 86*:104-112 (1980).

75. D. D. Wagner and R. O. Hynes, Fibronectin-coated beads are endocytosed by cells and align with microfilament bundles. *Exp. Cell Res. 140*:373-381 (1982).

76. L. VanDeWater, III, D. D. Wagner, E. B. Crenshaw, III, and R. O. Hynes, Fibronectin-dependent endocytosis by macrophage-like (P388D$_1$) and fibroblastic (NIL8) cells, in *Cellular Recognition* (W. A. Frazier, L. Glaser, and D. I. Gottlieb, eds.), Alan R. Liss, New York, 1982, pp. 869-878.

77. R. O. Hynes, Relationships between fibronectin and the cytoskeleton. *Cell Surf. Rev. 7*:97-136 (1981).

78. D. H. Bing, S. Almeda, H. Isliker, J. Lahav, and R. O. Hynes, Fibronectin binds to the Cl$_q$ component of complement, *Proc. Natl. Acad. Sci. USA 79*:4198-4201 (1982).

10

Wound Repair

Richard A. F. Clark
National Jewish Hospital
Denver, Colorado

Robert B. Colvin
Massachusetts General Hospital
Boston, Massachusetts

Over the past decade a sizable literature has documented an association of fibronectin with migrating and/or proliferating cell populations during embryogenesis, morphogenesis, wound healing, and tumor invasion. These findings, together with the in vitro data that demonstrate the capacity of fibronectin to bind cells to extracellular matrix and to mediate chemotaxis and perhaps contact guidance, have led to the hypothesis that fibronectin plays an essential role in multicellular organization.

Repair of injury is an essential adaptive response of multicellular organisms. Since repair mechanisms are in part regenerative, one would expect that these mechanisms share common components with developmental process. In addition, one would predict that generative and regenerative processes in diverse species would have common features since evolutionary pressure would select highly efficient and effective mechanisms of tissue organization and reorganization. We believe that fibronectin functions as one of these common, conserved features in intercellular organization because it can be demonstrated along the tracts of migrating cells during the development of such diverse species as the sea urchin [1], mouse [2], chicken [3], rat [4], and human [5] and during tissue repair in the mouse [6], rat [7, 8], guinea pig [9-11], rabbit [12-14], newt [15-17], and human [18].

Many reviews on various aspects of wound healing exist [19-23], including some consideration of fibronectin [24, 25]. Here we will attempt to provide some general concepts and hypotheses based on new information from this rapidly growing literature. We will review in some detail the evidence to date that suggests that fibroblasts, epithelial cells, endothelial cells, and macrophages are among those cells in wounds affected by fibronectin. We believe that fibronectin is a key component of the provisional matrix that promotes cell attachment, migration, and possibly differentiation. It should be emphasized, however, that formal proof of an essential role of fibronectin is lacking. Nonetheless, the known interactions of fibronectin with collagen, fibrin, glycosaminoglycans (GAG), and diverse cells, coupled with its conspicuous presence in all healing wounds analyzed to date, provide strong circumstantial evidence that fibronectin plays a major role in regenerative and repair processes.

First, we will briefly review the structural and functional relationships of fibronectin that have major implications for wound repair and outline the observed sequence of events in a prototype wound.

STRUCTURE-FUNCTION RELATIONSHIPS OF THE FIBRO-NECTIN MOLECULE

The morphological similarities of matrix fibronectin around cultured cells with that found in tissue give credence to the belief that interactions of fibronectin and cells elucidated in vitro have analogs in vivo. In vitro fibronectin binds to biological substrata, such as collagen or fibrinogen (fibrin), and simultaneously to cell surfaces and thereby promotes attachment, spreading, and growth

of those cells [26-29]. Fibronectin will facilitate cell attachment
to these matrices, whether derived from serum in the culture medium [27], added in the purified state to the substratum [28], or
synthesized and deposited by the cells in culture [29].

Fibronectin mediates adhesion of cells to substratum by cross-
linking the cells and substratum via specific binding domains that
exist along the primary structure of each of the two chains in the
440 kd dimer. These domains have been defined by the functional
properties of purified fragments obtained by limited proteolysis
and by selectively blocking the interaction of fibronectin and cells
or matrix molecules with polyclonal and monoclonal antibodies.
Thus, each chain of fibronectin binds to fibrin or fibrinogen via its
amino-terminal domain and through a site near the carboxyl terminus of the molecule [30-33], to collagen or gelatin via a 40 kd
domain near, but not at, the amino-terminal end [34-38], and to
cell surfaces through a 15 kd cell binding site that is approximately two-thirds the linear distance toward the carboxy terminus
[32, 39, 40].

Glycosaminoglycans (GAG), especially heparan sulfates that
are closely associated with the cell surface and occasionally intercalated in the cell membrane via their proteoglycan protein core
[41], are a likely binding site on the cell surface for fibronectin
[42,43]. In vivo GAG are thought to stabilize cell-substratum interactions mediated by fibronectin. Two binding sites exist for
heparin, the GAG most often used for in vitro studies, along each
chain of fibronectin. One site is located in the amino-terminal domain of the molecule [31, 33]. The interaction between this site
and heparin is relatively weak as it can be inhibited by physiological concentrations of calcium [44] or by 0.25 M NaCl [45] and,
consequently, has doubtful biological significances. A second heparin binding site is located close to the carboxyl terminus of fibronectin [31, 32]. Its heparin binding is less sensitive to salt [45]
and not sensitive to divalent cations [44], and therefore this site is
probably the primary region of GAG binding domain in vivo.
Fibronectin can also bind to hyaluronic acid, a nonsulfated GAG
[46-48]. However, fibronectin must exist as an aggregate to bind
efficiently to hyaluronic acid. Nonaggregated plasma fibronectin,
derived primarily from hepatocytes [49-51], when coupled to affinity columns, binds poorly to hyaluronic acid [52]. In contrast,

fibronectin from other cell types is normally aggregated and binds to hyaluronic acid with moderately high affinity [47].

Many of the interactions between fibronectin and other extracellular matrix molecules can be stabilized by the clotting factor XIII (fibrin-stabilizing factor). Factor XIII is a transglutaminase that enzymatically forms intermolecular γ-glutamyl-ϵ-lysyl covalent bonds between fibronectin and fibrin in clotting plasma [53], between fibronectin and collagen [54], and between two or more fibronectin molecules [55]. Fibronectin cross-bonded to fibrin(ogen) by factor XIII provides a much more adhesive surface for fibroblasts in vitro than does fibronectin alone or fibronectin with fibrin(ogen) but without the covalent bond [56]. Thus, fibronectin cross-linked by factor XIII to fibrin, to itself, or to collagen may be a more stable substrate in vivo for cell adhesion, migration, or proliferation. Interestingly, the genetic absence of factor XIII is known to be associated with an incompletely characterized defect in wound healing [57].

Finally, each subunit of fibronectin contains at least one sulfhydryl group [58, 59] located 170 kd from the amino terminus and probably a second farther toward the carboxyl terminus [60]. If these sulfhydryl groups are alkylated, the binding of fibronectin into the cell surface matrix is inhibited [58, 60]. Thus, besides the primary domains on each fibronectin monomer that bind to cell receptors and to extracellular matrix proteins, secondary interactions with the GAG and covalent cross-links mediated either by factor XIII or sulfhydryl groups can stabilize fibronectin-mediated cell-matrix associations. In addition, the fibronectin molecule is flexible and may exist in either elongated or folded forms in vivo, depending on solution or suface effects [61-63]. This structural pleomorphism may contribute to the molecule's biological interactions.

SEQUENCE OF EVENTS IN REPAIR OF SKIN WOUNDS

The initial event after an incisional or excisional wound is leakage of whole blood from torn vessels and rapid activation of the clotting system. The resultant coagulum fills the wound space in minutes with an adhesive gel that contains fibrin, fibronectin, platelets, red cells, and leukocytes and forms a temporary seal.

Within minutes to hours after the injury, additional neutrophils
and monocytes emigrate from nearby venules into the clot and adjacent tissue. The monocytes differentiate into macrophages that
debride the devitalized and denatured tissue. These cells reach
their peak at 2–3 days postinjury; however, their continued presence is necessary for wound repair. In fact, macrophages along
with fibroblasts and new blood vessels are thought to form an essential wound healing unit.

Within the first 24 hr the epithelium at the wound margin
elongates and flattens to form a noncornified tongue of cells, two
to three layers thick, that begins to dissect beneath the clot and
dessicated tissue. Cell division commences by the second day after
injury and occurs spatially behind but immediately adjacent to the
epithelial tongue. The epithelium migrates over a thick mat of fibrin and fibronectin that appears apart in both time and location
from the original clot. The basement membrane re-forms just behind the epithelial tongue at about the same site as epithelial proliferation. Although the leading epithelial edge migrates at a rate
of only 0.3 mm/day, concomitant wound contraction halves the
distance between wound edges. Thus, this process of re-epithelial
zation can completely cover a small (4 mm) defect in approximately 7 days.

Blood vessels, mostly at the wound base, sprout branches that
migrate into the wound bed as early as 2–3 days. As in the case of
the epithelial division, endothelial cell division occurs proximal
to the migrating cells. Thus, proliferation is initially concentrated
in tissue adjacent to the wound and occurs in association with increased endothelial fibronectin.

Fibroblasts begin to invade the wound bed by 2–3 days after
injury, along with macrophages and new capillary sprouts. By day
3 or 4 newly synthesized GAG, fibronectin, and collagen, all of
which most likely derive from the infiltrating fibroblasts, can be
detected within the wound. During this very early time period, fibronectin, hyaluronic acid, and types I and III collagen become
the major constituents of newly formed loose extracellular matrix.
As the wound space becomes filled with macrophages, microvasculature, and fibroblasts embedded within the loose extracellular matrix, the eschar is pushed outward and the migrating epithelium is
deflected over the surface of this so-called granulation tissue. By 5
days the fibroblasts residing in the upper portions of the granula-

tion tissue become oriented in a horizontal plane, interconnected by fibronectin and gap junctions and rich in actin microfilaments through which they mount a strong contractile force. The intercellular connections facilitate a concerted contraction that results in a reduction of wound size. Considerable remodeling (degradation, resynthesis, and realignment of fibrils) then occurs over the next 2-4 weeks. Hyaluronic acid is partially replaced by chondroitin sulfate-containing proteoglycans. Fibrin, fibronectin, and type III collagen are reabsorbed, and large bundles of type I collagen appear. Over the ensuing months tensile strength is restored toward normal and vascularity diminishes.

In the following paragraphs we will examine these aspects of wound healing in more detail, with a central focus on possible roles for fibronectin.

CLOT FORMATION AND HEMOSTASIS

An initial event after severe injury to any organ is leakage of whole blood from transected vessels and/or transudation of plasma from injured or reactive vessels. In either case fibrinogen, fibronectin, and other clotting factors, such as prothrombin and the plasma transglutaminase, factor XIII, bathe the wounded area. Activation of the clotting system by exposure of platelets to fibrillar collagen (types I, II, or III) and by release of tissue activation factors quickly causes an extravascular gel to form. This gel is created by the polymerization of monomeric fibrin and contains fibronectin whether the injury is an excisional wound [9] or a intense inflammatory reaction [64]. The fibrin and fibrinogen derive from both plasma and platelets (see Chap. 8) when whole blood is extravasated and from the plasma alone when a transudate accumulates from leaky blood vessels. The gel initially serves hemostatic and adhesive functions but soon thereafter probably serves an equally important role as the matrix for cell migration and as a source of chemical mediators that affect cell activity.

This extravascular gel is an important early component of wound repair, as judged by impaired wound healing in animals depleted of fibrinogen. Depletion of plasma fibrinogen can be accomplished by intravenous injection of ancrod, a snake venom enzyme. Ancrod cleaves fibrinopeptide A from fibrinogen, which is

then rapidly cleared from the circulation by the reticuloendothelial system [65]. Such treatment does not deplete plasma fibronectin [66] or activate factor XIII. Acute depletion of fibrinogen
in dogs with low doses of ancrod caused dehiscence of skin wounds
within 3 days without recurrence of gross bleeding [67]. In a series of studies in the rabbit, Brandstedt and his associates [68-70]
found that systemic defibrination with ancrod caused skin incisions to have decreased tensile strength and sponge implants to
have decreased deposition of a fibrillar matrix (probably fibrin-
fibronectin), less fibroblast accumulation, and diminished collagen
synthesis. These observations suggest that fibrin is necessary for
fibroblast migration and matrix formation.

Fibrin probably acts in vivo by forming a lattice that incorp-
orates fibronectin. Circulating plasma [^{125}I] fibronectin concentrates in rat abdominal incisions within 15-30 min of wounding
[71]; fibronectin appears by immunofluorescence in the extravascular clot of guinea pig skin wound by 3 hr after injury [9]; material presumed to be fibronectin, which agglutinates gelatinized
particles, drains from deep wounds in patients at levels exceeding
the plasma [72]; and plasma fibronectin concentrates in delayed
hypersensitivity skin test sites with fibrin as determined by immunofluorescence and radioisotopic techniques [64]. We have
demonstrated, in fact, that the accumulation of significant fibrillar
fibronectin in delayed hypersensitivity reactions is dependent on
the deposition of fibrin since a patient with afbrinogenemia failed
to deposit either fibrin or a reticular network of fibronectin in
the interstitium of a positive skin test site [73]. Thus, the presence of a fibrin gel in a wound site is probably indicative that
reticular fibronectin is also initially present.

Wound dehiscence is also a primary manifestation of certain
patients with cogenital dysfibrinogenemias, namely, the Paris I
and Cleveland I types [74]. Curiously, abnormal wound healing is
not an obvious feature of most other dysfibrinogenemia or afibrinogenemia syndromes; however, any deleterious effect on healing
would be obscured by the necessary treatment of wound hemostasis with fibrinogen and compressive bandages.

Potent chemical mediators ("wound hormones") that attract
and activate the appropriate cells most likely emanate from the
initial wound gel. Indeed, implanted homologous plasma clots are

readily invaded by fibroblasts and newly formed blood vessels over several days [75]. Products released from the clot gel by thrombin, plasmin, or other proteases may act as chemotactic factors. Certain large fragments of plasma fibronectin are chemotactic for monocytes [76], endothelial cells [77], and fibroblasts [78, 78a], and possibly fibronectin itself is chemotactic for endothelial cells and fibroblasts. In addition, the fibronectin matrix may act as a contact guidance system for any or all of these cells. Fibrinopeptide B, which thrombin cleaves from fibrinogen, is chemotactic for neutrophils [79], and thrombin itself is a chemoattractant for monocytes [80]. Thrombin can stimulate fibroblast proliferation and fibronectin synthesis and release [81, 82], thus creating a positive feedback system. Fibronectin itself has also been recently shown to act as a growth factor for fibroblasts [83]. Platelets are another immediate source of fibrinogen and fibronectin, as well as preformed mediators, such as platelet-derived growth factor, which promotes fibroblast proliferation and chemotaxis [84], and platelet factor 4 which is chemotactic for monocytes [85]. Components of the kinin and complement system may also contribute chemotactic factors, but a discussion of these is beyond the scope of this review. In summary, the list of potential wound hormones quickly becomes exceedingly long.

Factor XIIIa from plasma or platelets is another likely participant in the early clot. Factor XIIIa is a transglutaminase that stabilizes fibrin by forming γ-γ and α_n cross-links between specific glutaminyl and lysyl groups [86]. Stabilization is also promoted by covalently linking of α_2-plasmin inhibitor to the fibrin gel [87]. Several lines of evidence support the view that factor XIIIa may exert an action on wound healing via fibronectin. First, factor XIIIa can cross-link fibronectin to fibrin, collagen, and itself [54, 55, 88]. Factor XIII-deficient plasma does not support fibroblast outgrowth from explants, a process involving a combination of migration and proliferation [89]; however, conflicting results are reported [86]. Fibroblasts adhere more readily in vitro to fibronectin, which is cross-linked to fibrin by factor XIIIa [56], and fibroblast proliferation in vitro has been reported to be stimulated by factor XIIIa [90].

Deficiency of plasma factor XIII characteristically results in delayed cessation of bleeding, spontaneous abortions, and about a 14% prevalence of defective wound healing [86]. The reasons for

the variable manifestation of poor healing are not known but may
be in part related to different genetic forms of the deficiency. Although cellular transglutaminase from macrophages [91] and fibroblasts [92] differ in substrate specificity from plasma transglutaminase [93], they could provide important local alternative sources
of the enzyme.

Experimental studies of factor XIII administration to animals
have been limited. Predictably, the effects have not been dramatic
since normal animals would have an ample supply of endogeneous
factor XIII. Biel et al. [94] reported that local treatment with partially purified factor XIII from placental extracts increased the tensile strength of skin wounds in guinea pigs. Increased granulation
tissue occurred in rabbits given local or systemic human factor
XIII [95]. Knoche and Schmidt [96] gave human factor XIII intravenously to rats with muscle wounds. They found increased incorporation of [^{3}H] thymidine in fibroblasts at 72 hr and more
oriented fibroblastic growth and vascularization at 7 days, compared with controls given human albumin or saline. Further studies are warranted to clarify the role of these enzymes in wound
healing.

MACROPHAGES

Macrophages participate in wound healing and are probably of
critical importance [97]. In addition to their traditional roles in
phagocytosis, enzymatic digestion of provisional matrix, and remodeling permanent matrix, macrophages are a potent source of
an angiogenic factor [98, 99], transglutaminase activity [100], interleukin 1, a stimulator of fibroblast growth [101], collagenase
[102], and fibronectin [103-105]. The interactions of macrophages with fibronectin are detailed by VandeWater in Chapter 9.
Only certain points relevant to wound healing will be emphasized
here.

Monocytes are attracted to fibronectin, at least in fragmented
form [76]. In addition, fibronectin has a chemokinetic effect on
macrophages. These activities are reminiscent of the action of collagen fragments on macrophages [106]. These findings suggest
that degradation of the matrix can signal more monocytes to emigrate from the circulation to the site of injury.

Fibronectin promotes attachment of macrophages to gelatin, collagen, or fibrinogen substrates [107, 108] and, in some assays, ingestion of gelatin-coated particles [109-113]. Godfrey and Purohit [114] have found that fibronectin, obtained as a product of antigen-stimulated lymph node cells, causes aggregation of guinea pig macrophages. Fibronectin is necessary for optimal immobilization of macrophages in the migration inhibition assay [115] (Colvin et al., unpublished). Thus fibronectin may serve as a matrix for macrophage adherence, aggregation, and retention at sites of inflammation and, in addition, act as an opsonin in vivo. From its known in vitro binding properties, opsonization of collagen (especially denatured forms), fibrin, DNA, actin, complement-coated particles (Clq, alternative pathway), and certain bacteria might be predicted, although such a role in vivo has not yet been demonstrated.

Some in vivo correlates of these observations have been obtained. Fibronectin occurs around mononuclear cells in wounds [9] and around many HLA-DR-positive cells, some of which react with monocyte-specific antibodies, in the perivascular infiltrate of delayed hypersensitivity reactions [73].

In addition, surface fibronectin is present on inflammatory macrophages in the peritoneal cavity in association with fibrin-fibrinogen [116, 117]. The latter phenomenon indicates that fibronectin-macrophage binding does occur in vivo and does not require exogenous hepain, as do some in vitro phagocytic assays.

The macrophage responds to fibronectin in other ways. Plating monocytes on fibronectin for a few hours increases Fc and C3 receptor activity, suggesting a major alteration in the state of the plasma membrane [107]. Fibronectin induces an increase in monocyte growth factor activity, which may in turn greatly affect other cells [98, 99].

RE-EPITHELIALIZATION

Epithelial Basement Membrane

The normal anchoring mechanism of epithelium to the underlying extracellular matrix is the basement membrane. Since the introduction of the electron microscope, the term "basement membrane" is used specifically to refer to the continuous sheet of extracellular matrix material that forms electron-lucent and electron-dense layers at the interface between tissue cells, such as epithelium, endothelium, and smooth muscle cells and extracellular matrix. These specialized matrix structures appear as bands by electron microscopy because the tissue is usually transected for specimen preparation. The electron-lucent bands are called the lamina lucida interna (near the epithelium) and externa, and the interposed, single electron-dense band is termed lamina densa. Basement membrane matrices are products of the adjacent cells and are composed of proteoglycans, collagen, and noncollagenous glycoproteins. More specifically, most basement membrane contain fibronectin, laminin, type IV collagen, and heparan sulfate proteoglycan, as well as smaller quantities of other glycoproteins and proteoglycans.

In the epidermis, the plasma membranes of basal cells, those cells that reside directly on the basement membrane, are highly specialized and sometimes are considered to be an intrinsic part of the basement membrane. One such specialized structure of the basal cell plasma membrane is hemidesmosomes. These are thickened plaques in the plasma membrane into which keratin fibers or tonofilaments insert on the cytoplasmic side (Fig. 1) [18] and from which anchoring filaments run from the external side to insert into the lamina densa [119]. The molecular nature of hemidesmosomes and anchoring filaments has not been elucidated at this time.

Another component of basement membranes specific for stratified squamous epithelium and an intrinsic part of the basal cell plasma membrane is bullous pemphigoid antigen (BPA) (see Fig. 1). This glycoprotein was identified as the target of autoantibodies in patients with a blistering skin disease called bullous pemphigoid in which the epidermis detaches from the underlying lamina densa. Such antibodies to BPA also cause detachment of epidermal cells in vitro [120]. Ultrastructural studies have localized BPA to the lamina lucida, the electron-lucent region adjacent to the cell membrane [121, 122]. Stanley et al. [123], using BPA antiserum, have immunoprecipitated a protein from cultured mouse and human epidermal cells that migrates in sodium dodecyl sulfate polyacrylamide gel electrophoresis (SDS-PAGE) as a 220 kd subunit polypeptide after sulfhydryl reduction and as a larger complex without reduction. In contrast, Diaz et al. [124, 125] have isolated a pro-

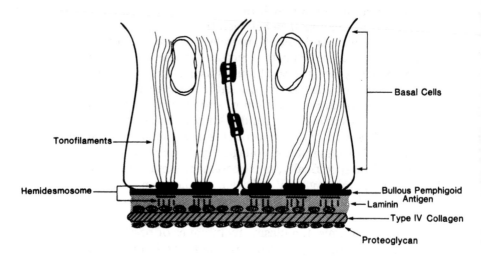

Figure 1 Schematic diagram of the normal epidermal-dermal interface. During the migrating phase of wound re-epithelialization the structures depicted
below bullous pemphigoid antigen are replaced by a provisional matrix of fibronectin and fibrin.

tein (18,000–20,000 kd) from human urine and skin that reacts
with bullous pemphigoid antibodies. The relationship of these two
proteins is not clear at the present time. Although the biological
function of BPA is not known, that BPA antibodies bind to the
epidermal basement membrane and subepidermal blisters form in
patients with bullous pemphigoid suggests that BPA may be involved in epidermal cell adhesion.

    Although fibronectin has been localized by immunohistochem-
ical techniques in the basement membrane matrix around smooth
and striated muscle, and blood vessels and thyroid, breast, and gut
epithelium of normal mammalian adults [126–129], the presence
of fibronectin in stratified squamous epithelial basement membranes has been less clear. In the corneal epithelial basement membranes of the guinea pig, rabbit and human, no fibronectin is detectable [13, 130]. In the epidermal basement membrane, some
studies find little or no fibronectin in humans [126], guinea pigs
[10], mice, or rats [131], but others describe fibronectin in this
location in guinea pigs [9], rats [129, 132], and variably in the

human [133]. In the study by Couchman et al. [129], patchy, thin bands were present in the lamina lucida externa by immuno-electron microscopy that were not well seen by immunofluores-cence. Fibronectin in glomerular basement membrane has been controversial for some time, despite careful study by several groups [134]. In contrast, fibronectin is a prominent feature of embryonic basement membranes [3], including rat [132], rabbit [135], and human (Tonnesen and Clark, personal observations) stratified epithelial basement membranes.

Several factors could contribute to the variation in basement membrane fibronectin, including differences in sensitivity of the techniques, differences in species or age, the presence of normal tissue remodeling, and past injury. If basement membrane fibro-nectin in adult stratified squamous epithelium were a remnant of epithelial repair processes, this might explain why corneal base-ment membrane fibronectin increases with age in humans [136] and why it is variable in skin and glomerular basement membranes. The presence of fibronectin in hair follicle basement membrane has been attributed to remodeling that occurs in the normal cycle of hair growth [129]. The gut basement membrane is a special case, since it is one of the few sites in which *lateral* migration of the epithelial cells on basement membrane is a normal process in the adult. Taken together, the data support the view that fibronec-tin is a major basement membrane component primarily during embryogenesis and regenerative processes. In adults its presence may reflect subliminal repair and/or migration.

All basement membranes are enriched in noncollagenous gly-coproteins even if they contain BPA or fibronectin. Laminin is such a glycoprotein and occurs in fairly large quantities in all base-ment membranes examined [137]. This 1 million molecular weight, asymmetric molecule with three 200 kd chains at one end and a 400 kd chain at the other has been isolated from a variety of carcinoma-derived cell lines [138, 139] and digests of human kid-ney and placenta [140]. Although most studies [141], including those on corneal epithelium [14], agree that the bulk of laminin lies beneath BPA (Fig. 1) and that a plane of mechanical weakness exists between the BPA and laminin, it is not clear whether the majority of laminin resides in the lamina lucida [137, 142, 143] or in the lamina densa [144, 145]. Nevertheless, laminin is be-lieved to be important in stable epidermal cell-substrate interactions. Several models have been proposed [146] based on in vitro evidence that laminin promotes attachment of epithelial cells to type IV collagen but not to other collagen types [147, 148] and that soluble laminin stabilizes corneal epithelial cells [149]. One attractive current view is that laminin forms a bridge between type IV collagen in the lamina densa and the cell at the basement membrane interface.

Entactin, a newly described sulfated glycoprotein of 158 kd, is another noncollagen glycoprotein that has been identified in rat and mouse basement membranes with a similar, but not identical, distribution to laminin [150]. Its role in the basement membrane and its participation in wound healing have not been defined.

Proteoglycans are present in basement membranes, typically visualized in the lamina lucida as regularly spaced negatively charged groups with ruthenium red or other cationic dyes (Fig. 1). Proteoglycans from various basement membranes are quite distinct as exemplified by the heparan sulfate proteoglycan isolated from the EHS tumor matrix, a large macromolecule (750 kd) with roughly eqeal proportions of heparan sulfate (five 70 kd chains) and protein [151]; and the heparan sulfate proteoglycan (120 kd) from glomeruli with heparan sulfate chains of 18 kd [152]. The proteoglycans probably add stability to the basement membrane cell complex by binding to several of the other constituents of basement membrane [146]. In addition, heparan sulfate forms a negatively charged shield that blocks passage of anionic macromolecules through the basement membrane, thus acting as a filter [153, 154].

All basement membranes, with the possible exception of rodent and chicken cornea [14, 155], contain type IV collagen. This collagen has been localized to the lamina densa of basement membranes (Fig. 1) [156] and is distinguished by its ability to form a continous sheet of extensible meshwork rather than periodic banded fibers [157]. The interstitial collagens (types I, II, and III), which form cross-banded fibrillar structures in vivo, have not been found in basement membranes [158]. Another nonfibrillar collagen (type V), which forms pericellular matrices, occurs in codistribution with type IV in the mesangial matrix and basement membranes of the kidney [159]. Type V collagen has been found at the dermoepidermal junction by some investigators [160] but not by others [161]. As will be discussed, type IV collagen is a good substrate for epidermal cell attachment [147]. Type V collagen may also play a role in cell adhesion [162].

Provisional Substrate

A critical process in epithelial repair is the rapid migration of epithelial cells over the wound surface to close the defect and re-establish an effective barrier between the host and the environment. If the epidermis is split off the basal lamina by suction and the epidermal roof of the blister remains intact, epithelial cells from hair shafts within the blister base migrate over the residual basement membrane and effectively close the wound within 24 hr of injury [163]. If, however, the epidermal roof is removed and the underlying basement membrane and dermal structures are allowed to dessicate, as in an open wound, the epithelium must dissect under the dried, denatured tissue. In these situations re-epithelialization occurs much more slowly, for example, 3 days for open blisters and 7-9 days for 4 mm excisional wounds. The substrate for migration in situations similar to these latter circumstances has not been completely defined, although type V collagen, BPA, fibrin, and fibronectin have been found immunohistochemically [10, 12, 15, 160, 164] and fibrin had been noted by electron microscopy [163, 165].

In all types of epidermal wound, the migrating epithelial cells develop substantial actin bundles in their peripheral cytoplasm [166] and retract their tonofilaments to a perinuclear position. Although they continue to be linked together with desmosomes, they have fewer such connections than normal epidermis. In the case of an intact suction blister the leading cells re-form complete hemidesmosomes soon after contacting the residual basal lamina [163]. In the open wounds, the leading cells appear to form hemidesmosome-like structures that come into intimate contact with extracellular fibrils presumed to be fibrin in ultrastructural studies using routine transmission electron microscopy [163, 165]. However, these structures do not have a clear extracellular dense line parallel to the thickened cell membrane, as do classic hemidesmosomes [167]. In addition, tonofilaments do not insert into the intracellular plasma membrane thickenings as they do in typical hemidesmosomes [167]. These plasma membrane structures seem more analogous to the intermediate densities [168] or the fibronexus-like epinexus [169] observed in basal cell epithelioma.

Regardless of the exact morphological and biochemical nature of the bonding between the leading epithelial cells and either the residual basement membrane or the fibrinlike material, junctions do occur between the cells and the substrate during epidermal migration, which suggests that the cells slide over one another to advance along the wound surface in both circumstances [163]. In the cornea, in contrast, the migrating cells do not form hemidesmosomes or other apparent junctions with the basement membrane until migration is complete, and thus the lead cells may continue movement [14]. The mechanistic difference between epidermal cell and corneal epithelial cell movement may explain in part why re-epithelialization of the cornea occurs 5-10 times faster than closure of the epidermis when the underlying basement membrane is removed. Of course, in re-epithelialization of open wounds, the epidermis must dissect a path through the connective tissue, separating dessicated tissue and debris from viable tissue, and complete re-epithelialization is slowed even further. The longer time needed for wound closure in this latter situation may be the result of the physical barrier, or it may be secondary to diversion of metabolic energy for degradation [170-172] and phagocytosis [163] of connective tissue in the path of the migrating epithelium.

In open skin wounds the determinants for the dividing line between the viable and nonviable tissue are not known; however, the epidermal cells seem to follow a pathway composed of dense irregular fibrillar material that condenses to form a thick band at the interface of the migrating epithelium with the dermis (Fig. 2) [10]. This material contains no detectable BPA, laminin, or type IV collagen [10], but fibronectin and fibrin (Fig. 3), [10] are invariably present by immunofluorescence technique. Repesh et al. [12] have also reported that rabbit epidermal cells migrate on fibronectin and fibrin filaments. These findings are consistent with previous ultrastructural data that the migrating epithelial cells had no basement membrane and were often seen in contact with a fibrinlike fibrillar material [163, 165]. From our own studies it is clear that these cells do not migrate over the original fibrin clot that forms the eschar that is ultimately dissected out of the wound.

Plasma and platelets are the only known sources of fibrinogen; thus, the fibrin in this newly formed fibronectin-fibrin pro-

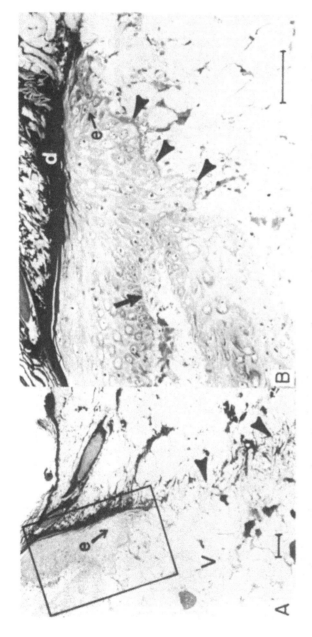

Figure 2   (A) Epon section (1 μm) of the edge of a 3 day excisional wound in guinea pig skin. The migrating epithelial tongue (e) dissects between the necrotic tissue containing inflammatory cells (above) and the viable dermis (V). Fibrin strands are interposed between collagen bundles (arrowheads). Bar = 5-μm. (B) Higher magnification of the same section. Beneath the migrating epidermis (e) an irregular thickening of the basement membrane zone (BMZ) (arrowheads) is interconnected with fibrillar material deposited in the underlying dermis. This can be contrasted to histologically appearing normal BMZ beneath stationary epidermis (arrow). Dense fibrin eschar, postsyneresis (d), lies above the migrating epidermis. Bar = 50 μm. (From Ref. 10.)

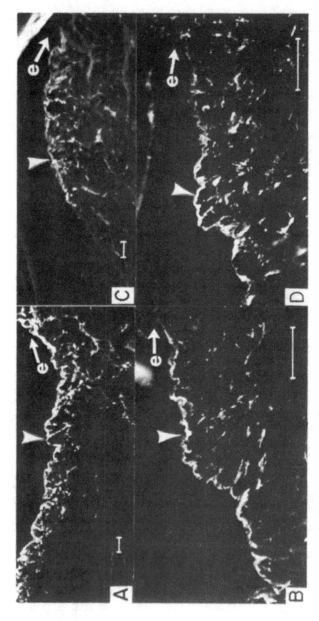

Figure 3 Specimens of 3 day excisional wounds in guinea pigs stained with antifibrinogen antibodies (A,B) and antifibronectin antibodies (C,D). (A,B) Fibrin is disposed beneath the migrating epithelium (e) in an irregularly thick linear fashion along the BMZ and into the adjacent dermis (arrowhead). (C,D). Fibronectin also appeared along the BMZ of the migrating epidermis (e) as a thick, irregular band with strands extending perpendicularly into the dermis (arrowhead). Arrows indicate tip of migrating epidermis. Bars = 50 μm. (Modified from Ref. 10.)

visonal matrix presumably is derived from the circulation. This
possibility is given credence by the fact that the blood vessels near
the migrating epidermis continue to show marked permeability to
carbon particles throughout the re-epithelialization process [131].
The fibronectin in this provisional matrix, in contrast, could be
either plasma derived or synthesized locally. In fact, examination
of wounds placed in a well-healed rat skin graft on immunosup-
pressed mice with reciprocal species-specific antibodies to rat and
mouse fibronectin showed that the fibronectin was mostly plasma-
derived, mouse fibronectin, during the first 4 days after injury
and, conversely, mostly in situ produced, rat fibronectin, at later
times in the re-epithelialization process [131]. The consistent presence of locally derived fibronectin beyond 4 days of wound repair correlates with the time that fibroblasts grow into the wound
defect from the base and sides to form granulation tissue [131].
These cells are probably one source of locally produced fibronectin.
    Keratinocytes can also produce fibronectin in vitro [4, 173–
176], thus raising the possibility that the migrating epithelial cells
themselves contribute to their substrate fibronectin. Synthesis of
fibronectin by normal human cervical epithelial cells in vitro has
been related to a relatively "dedifferentiated" state, as judged by
the lack of keratin and the presence of fibronectin in the cells at
the margins of the cell clusters [177], perhaps analogous to the
wound epidermis. Epidermal cell fibronectin synthesis in vivo is
also suggested by cytoplasmic fibronectin in basal epidermal cells
in regenerating newt limbs [15], intranodular fibronectin of basal
cell epithelioma [178, 179], and intraepidermal fibronectin in
human fetal skin (Tonnesen and Clark, unpublished data). In any
case, the binding of both plasma-derived and locally produced fibronectin under the migrating epithelium is consistent with in
vitro data that plasma and cellular fibronectin can produce extracellular matrices in concert [180].
    In spite of the absence of ultrastructurally obvious basement
membrane, during epidermal migration fluorescein-conjugated antibodies to BPA do stain the basal side of the migrating cells all the
way out to the extended pseudopodia of the leading cell [10,
181]. The significance of this finding is unknown but may indicate that BPA in conjunction with fibronectin and fibrin plays a
role in the adherence-migration process of re-epithelialization. As

expected from the absence of basement membrane on electron microscopy, laminin and type IV collagen do not occur under the migrating epithelium [10, 160, 164, 181]. After the epidermal defect is repaired and epidermal migration has ceased, the basement membrane is reestablished under the epithelial cells [165]. In our own studies on open guinea pig wounds, we observed, as the defect is repaired, type IV collagen and laminin reappeared approximately 10-20 cells behind the tip of the migrating as fibrin and fibronectin gradually disappeared [10]. Thus the reestablishment of basement membrane and the loss of the provisional matrix occurs in a zipperlike fashion from the original wound margin toward the center until the entire wound surface is re-epithelialized.

To determine whether fibronectin and fibrin are the substrate for epithelial migration in a "pure" epithelial wound, healing superficial rabbit corneal epithelial wounds have been examined, similar to mild human corneal abrasions [13, 14]. In the normal cornea, fibronectin was not detected in the epithelial basement membrane. Shortly after wounding (8 hr), fibronectin deposited on the denuded corneal surface as continuous, prominent layer. The epithelium had begun to migrate over the deposited fibronectin by 22 hr and by 52 hr had completely recovered the denuded surface. Similar fibronectin deposits have been noted under the wounded corneal epithelium by others [130]. Fibrinogen-fibrin was also present on the initial base wound surface (1 hr) and preceded the fibronectin in maximal intensity. Once the wound was re-epithelialized, the subepithelial fibronectin and fibrin layer then progressively disappeared. The migrating epithelial cells lacked hemidesmosomes, which were not regenerated until after migration was complete (1 week) [14].

Fibronectin in these corneal wounds had an inverse relationship to BPA (Fig. 4). The BPA was removed with the epithelium and only reappeared after wound re-epithelialization was complete and as fibronectin diminished, unlike the situation in the skin. Laminin, detectable in the normal corneal basement membrane, was not removed by the epithelial scrape wounds, indicating that the basement membrane is split between the laminin and BPA layers [14]. After superficial keratectomy, however, laminin was removed and reappeared under the migrating epithelium for a variable distance, but was absent from the migrating edge. Fibronectin

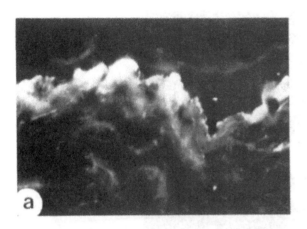

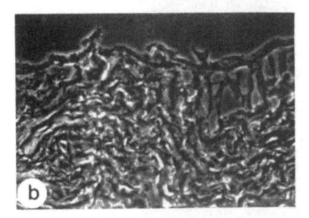

Figure 4   Corneal wounds stained for fibronectin (a, d, e) or fibrin-fibrinogen
(c). Superficial keratectomy at 48 hr (a, b, c), and at 2 weeks (e) and scrape
wound at 22 hr (d). The fibronectin is a fine band on the scrape surface (d)
but extends more diffusely into the bare stroma (a), which is superficially
coated with fibrin-fibrinogen (c). The epithelium migrates across this surface
and can be seen on right in (b), a phase micrograph of same field as (a). After
the epithelium covers the surface in 2–3 days, fibronectin accumulates in the
stroma around the proliferating fibroblasts, which fill in the stroma defect
over the next several weeks (e). a, b, c, ×230; d, e ×92. (From Ref. 14.)

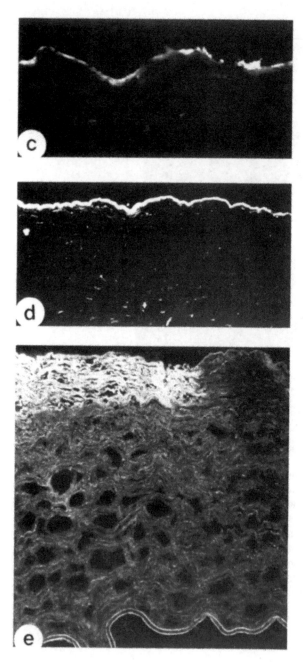

Figure 4 (continued)

is deposited as a more diffuse, irregular band on the bare stroma after keratectomy. Curiously, type IV collagen was not detectable in rabbit corneas before or after wounding (sometimes trace amounts were found in keratectomy wounds at 2-4 weeks). The corneal basement membrane had no detectable type IV collagen (rabbit or GP) or less than the conjunctival basement membrane (human) [14, 136]. Since this would be a unique absence of type IV collagen in a basement membrane, the antigenic epitopes may be merely obscured or altered. Further work is required to elucidate the reason for this finding. Type IV collagen has been reported by others in human corneas [182].

The presence of fibronectin beneath migrating corneal epithelium as well as under migrating epidermis suggests that fibronectin, together with fibrin, forms a provisional matrix that promotes migration and adhesion of either type of epithelial cell. The fibrin and fibronectin seem to correspond to the irregular fibrillar material haphazardly deposited on the lamina densa or bare stroma when immunofluorescence studies are compared to plastic 1 μm sections taken from the same tissue. Although fibrin fibrils with characteristic periodicity have been described under the migrating epidermis [163, 165], immunoelectron microscopic studies are required to determine the ultrastructural relationship between the fibronectin-fibrin and the migrating cells.

The presence of fibronectin in the epithelial basement membrane zone is a common feature of wound repair and embryonic development [183]. Rabbit fetal corneas at 13 days' gestational age have fibronectin between the lens and surface epithelium [135]. A separate delicate line of fibronectin appears under the coreneal epithelium at 29 days, preceding the growth of epithelial cell layers, which occurs after birth. Subepithelial fibronectin disappears and is no longer present in young adult rabbits. Fibronectin in the corneal basement membrane precedes laminin accumulation, which becomes detectable after birth in the guinea pig (Golub and Colvin, unpublished). Likewise, fibronectin occurs in a linear fashion at the dermoepidermal junction of human fetal skin examined during the second trimester (Tonnesen and Clark, personal observations). Thus, wound repair recapitulates ontogeny with respect to fibronectin but differs in that fibrin accompanies the fibronectin deposition during healing.

In summary, during epithelial migration in skin and deep corneal wound, the permanent, normal basement membrane components, type IV collagen and laminin are not detectable, nor is BPA present in superficial corneal wounds. The only substrates uniformly present under the migrating epithelium, in all types of wounds in the skin and cornea, were fibronectin and fibrin, neither a normal component of these basement membranes (Figs. 3 and 4). With time (7–14 days), this provisional substrate is removed as permanent components are synthesized [10, 13].

Interaction of Epithelial Cells with Fibronectin

Fibronectin substrates enhance the migration of fibroblasts in vitro [184], presumably by increasing adherence to the substratum. This seems to apply to epithelial cells as well [17]. However, there are conflicting data on whether fibronectin does [185, 186] or does not [147, 187, 188] promote epithelial cell attachment; in the last three studies, epithelial cells attached to laminin in preference to fibronectin in vitro, as did carcinoma cells [189]. These studies have used fibronectin alone or fibronectin on gelatin or collagen. Fibronectin binding by epithelium was shown with intestinal epithelial cells isolated with hyaluronidase [185] and with a primary epithelial culture [186], possibly resembling the more "de-differentiated" migrating epithelial cell. Fibronectin in solution diminished basal blebs of corneal epithelium resting on a membrane filter, as did type IV collagen and laminin [149], clearly indicating a cellular recognition response. Some of the variation in results of epithelial fibronectin adherence may be related to the state of differentiation, the conditions of isolation of the epithelial cells, or the form of the fibronectin.

Recent studies on the adherence of human keratinocytes to various substrates support the notion that cellular differentiation greatly affects cell adherence to extracellular matrix proteins [190]. Human neonatal foreskin keratinocytes grown in 0.1 mM Ca^{2+} contain tonofilaments in a perinuclear disposition, have few intercellular desmosomes, and do not stratify, features shared with migrating epithelium during wound repair. In assays using either primary isolates or second passaged cells, the keratinocytes in 0.1 mM Ca^{2+}, and serum-free medium adhered similarly to laminin, type I and III collagen, and fibronectin. This adherence seemed specific for these extracellular matrix proteins since only one-quarter to one-third the number of cells attached to fibrinogen or albumin. Adherence to type IV collagen was intermediate. Similar results were found when guinea pig epidermal cell spreading was assayed on these same protein coats [191]. From these data, both groups of authors surmise that keratinocytes, phenotypically similar to those cells in the migrating epidermis, attach as well to non-basement membrane matrix proteins as to the basement membrane protein laminin. The broad specificity would allow the epidermal cells during reepithelialization to migrate on the available wound surface substrate whether the wound was open, closed, deep, or shallow.

Ex vivo studies of cell migration from explants or wound margins are likely to provide instructive analogies with wound healing. In one detailed analysis by Billig et al. [118], epithelial cells from chick embryo corneas were cultured on gelatin substrates. Cells at the perimeter of each cluster had well-spread lamellae with radially oriented focal contacts, fibronectin, and actin at their leading edge. Tubulin but not prekeratin extended to the cell periphery. Laminin was not detectable, but particulate fibronectin was found on the gelatin substrate away from the cell margin. Removal of fibronectin from the fetal calf serum in the medium prevented the deposition of fibronectin, but the cells had a normal morphology. The authors suggest that exogenous fibronectin is deposited on the substrate and reorganized by the peripheral cells as they migrate. Such a mechanism would fit well with the observation of epithelial healing in vitro in which cells migrate over a layer of fibronectin initially from an exogenous (plasma) source. The importance of both fibrinogen and fibronectin for epithelial wound closure is supported by the elegant studies of Donaldson and Mahan [17], in which pieces of glass coverslip coated with optimal concentrations of fibronectin or fibrinogen and implanted under the margin of an adult newt skin wound were demonstrated to promote the same amount of epidermal migration as the wound bed itself. In contrast, pieces of coverslips coated with serum or albumin allowed little migration to occur. Finally, migration on fibrinogen- and fibronectin-coated substrate could be blocked by anti-fibrinogen and anti-fibronectin antibodies, respectively.

Migration of cells is a complex process, requiring both adhesion and controlled detachment as the cells move. The permanent basement membrane components may produce less reversible, firmer adhesion that is not optimal for cell movement. Although studies to date do not rule out participation of other plasma [192, 193] or cellular adhesive molecules [194], they do provide strong circumstantial evidence that fibronectin and fibrinogen are important, perhaps essential components of the substrate used by migrating epithelial cells in vivo. Since no fibronectin gradient is apparent on the epithelial substrate, the directionality of movement may be secondary to absence of contact inhibition on the margin of the wound or to other undefined factors.

NEOVASCULARIZATION

Considerable interest has focused recently on the mechanisms of control of blood vessel growth, particularly as related to wound healing and tumor growth [195]. Several cell extracts from tumors, the retina, or adipocytes have been shown to have angiogenic activity, that is, the ability to promote new capillary growth [196–199], but the active factor(s) in these cell extracts have not been isolated and are not necessarily related to the angiogenesis of wound healing. A number of proliferative and chemotactic factors have been described that may relate to angiogenesis of wound repair, including macrophage-derived growth factor [98], fibronectin [77], and heparin [200]. Recently, Banda et al. [201] have partially purified a small (2–14 kd) molecule from rabbit wound fluid that not only promotes capillary endothelial cell migration in vitro but also stimulates angiogenesis in the rabbit corneal implant assay. Interestingly, this molecule has no mitogenic activity. This activity profile actually corresponds to the initial independence of endothelial cell migration [202] and angiogenesis [203] from endothelial proliferation. However, prolonged angiogenesis is almost certainly a complex process that probably depends on at least three types of phenomena: an initial induction of angiogenesis by a factor such as the one described by Banda; a mitogenic stimulus for continued growth; and an appropriate extracellular matrix for endothelial cell migration or, more geometrically accurate, capillary bud formation and extension. We, as well as others [204], have concentrated our attention on this latter aspect of angiogenesis.

Correlation of Microvascular Fibronectin with Endothelial Cell Proliferation

In studies of 4 mm full-thickness skin wounds in the flanks of guinea pigs, we have shown that a transient increase in microvascular fibronectin (Fig. 5) accompanied the proliferation (Fig. 6) that occurred in vessels adjacent to wounds and in the neovasculature of the wounds [11]. Vessels within 0.5 mm of the wound edge initially had a normal staining intensity for fibronectin as observed with immunofluorescence techniques (Fig. 5A). By 3 days, shortly after the onset of endothelial cell proliferation (Fig. 6), however, scattered vessels near the wound edge exhibited a marked increase in fibronectin (Fig. 5B), and by 5-7 days, this change became marked and consistent (Figs. 5C-E). By 9 days, fibronectin immunoreactivity of these vessels returned to normal (Fig. 5F). Vessels more than 0.5 mm from the wound perimeter showed no change in fibronectin at any interval. Blood vessels did not stain for fibrinogen at any time after wounding, which we took to indicate that nonspecific absorption of plasma proteins did not account for the increased fibronectin staining. At the margins of the granulation tissue, the newly formed blood vessels were intensely positive for fibronectin; however, deeper into the granulation tissue it was difficult to distinguish blood vessels from the reticular fluorescent background attributed to fibroblasts.

Several investigators have demonstrated that proliferating endothelial cells synthesize and secrete fibronectin in vitro [205-207]. Most of the secreted fibronectin is, in fact, deposited between the cell monolayer and the culture dish, creating a form of "basement membrane" [207]. Recently, we have obtained evidence that proliferating blood vessels in wound produce fibronectin in situ [8]. As previously stated, small excisional wounds were made in rat xenografts that had been placed on thymectomized mice treated with anti-lymphocyte serum to prevent immunologic rejection. Specific antisera directed against either mouse or rat fibronectin allowed identification of the source of fibronectin deposited in vessels. We found that only rat fibronectin was detected in the walls of reactive wound vessels, indicating that the increased amounts of fibronectin seen in proliferating blood vessels were indeed synthesized locally rather than deposited from the plasma.

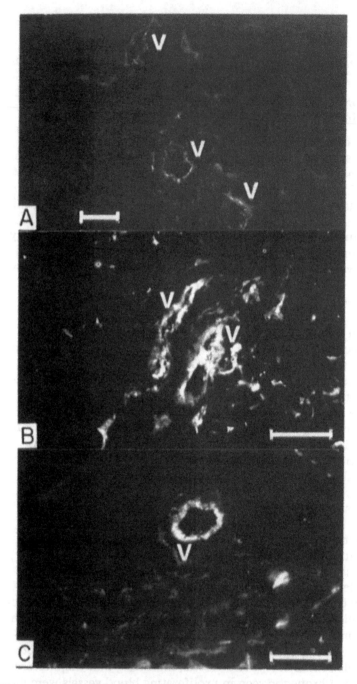

Figure 5 Immunofluorescence studies of small blood vessels (v) within 0.5 mm of an excisional wound in guinea pigs at various time periods after extirpation using fluorescein-conjugated anti-guinea pig fibronectin antibodies.

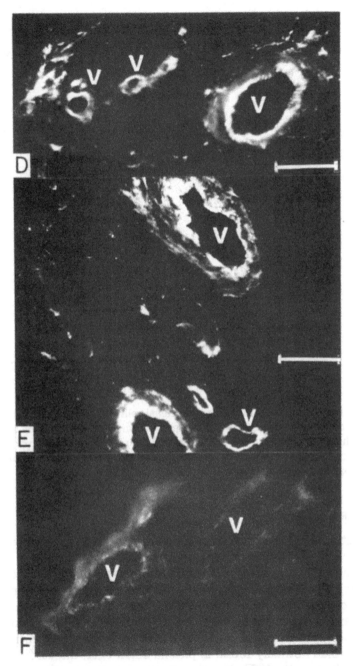

Figure 5 (continued)   (A) At day 1 vessels stain weakly. (B) Vessels at day 3 often stain brightly for fibronectin. (C–E) From days 5 to 7 vessels always have intense fluorescence. (F) By day 9 vessel fluorescence has returned to normal baseline. Bars = 50 μm. (From Ref. 11.)

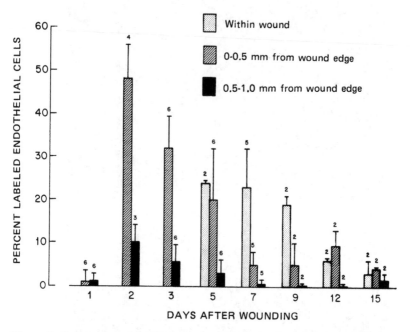

Figure 6 Percentage of proliferating endothelial cells 0–0.5 mm and 0.5–1.0 mm from the wound edge and within the granulation tissue as a function of time. Each bar, bracket, and number represents the mean ± standard error of the mean of percentage thymidine-labeled endothelial cells for n number of animals for the data point, respectively. (From Ref. 11.)

Role of Fibronectin

Recent studies suggest that fibronectin serves both as an adhesive substrate [208,209] and as a chemoattractant [77] for endothelial cells. Whether the directed migration is true chemotaxis or haptotaxis (contact guidance) needs to be addressed. In addition, it is not clear what form of fibronectin is most active. Although ample fibronectin might be expected to emanate from the wound due to local synthesis, it is puzzling that fibronectin in concentrations less than normally found in plasma might exert directed chemotactic activity [77]. If such a mechanism operates in vivo, obviously additional control mechanisms are required. An indirect mechanism by which fibronectin might promote angiogenesis has been advanced by Martin et al. [99]. These investigators found that monocytes plated on fibronectin developed increased activity of macrophage-derived growth factor that stimulates endothelial cell growth. Such an interaction is likely to occur in wounds since both macrophages and fibronectin are present.

A transient alteration of the subendothelial matrix might provide a more suitable substrate than basement membrane for endothelial movement and mitosis [210]. Recent studies [211] have indicated that the buds of newly formed blood vessels growing into a cornea in response to a thermal injury contained conspicuous fibronectin but no increase in factor VIII-related antigen, laminin, or type IV collagen. This response was seen as early as 6 hr after injury in the limbal vessels nearest the injury. Once migration ceased, the fibronectin was reduced. These in vivo observations are reminiscent of the "variant" bovine aortic endothelial cell detected by McAusland et al. [212, 213] in vitro, which produced greater quantities of fibronectin than normal endothelial cells and, in contrast, deposited fibronectin on the dorsal cell surface and in a trail when it migrated. According to their hypothesis, these variant cells are "pathfinders" that lay down a track of fibronectin for subsequent endothelial cells to follow, a novel follow-the-leader haptotaxis. Increases in vessel wall fibronectin are also associated with endothelial cell proliferation in cell-mediated immunological reactions [64] and may represent a general marker for a reactive (or reparative) process of endothelial cells. This alteration in the blood vessel basement membrane may provide a substrate more conducive for the migration of proliferating endothelial cells, may reduce the coagulative nature of the subendothelium, and may modulate the emigration of inflammatory cells (Fig. 7). Further studies are needed to elucidate these possibilities.

FIBROBLAST INFILTRATION AND WOUND CONTRACTION

Since fibronectin was originally observed in matrices of cultured fibroblasts and shown to promote adhesion of these cells to collagen, initial studies on fibronectin in wound healing were focused on the relationship between fibronectin and fibroblasts. Several investigations noted a spatial and kinetic correlation between fibroblast ingrowth and the appearance of fibronectin in granulation

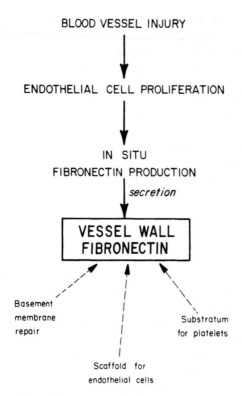

BLOOD VESSEL INJURY

ENDOTHELIAL CELL PROLIFERATION

IN SITU
FIBRONECTIN PRODUCTION
secretion

**VESSEL WALL
FIBRONECTIN**

Basement
membrane
repair

Substratum
for platelets

Scaffold for
endothelial cells

Figure 7 Sequence of events resulting in and proposed consequences of increased blood vessel wall fibronectin.

tissue (Fig. 8) [6, 7, 9, 11, 12]. Although quantitative studies are not available, granulation tissue fibronectin peaks at about the same time as maximal fibroblast proliferation and before the peak of newly formed collagen, as judged by immunohistochemistry studies of sponge implants [6].

As discussed, the initial source of extravascular fibronectin in wounds is plasma [71, 131] and possibly platelets (see Chap. 8). As the eschar, which contains fibrin and fibronectin as well as other clotting proteins and tissue debris, is dissected out of the wound by the migrating epidermis [10], continued increased vasopermeability of the blood vessels under the re-epithelializing epidermal cells and at the wound base leads to further accumulation of fibrin and fibronectin in the early wound site [131]. The resultant interstitial fibronectin under the migrating epithelium (Fig. 3) and at the wound base [9] is deposited in a haphazard coarse reticular pattern. In contrast, as fibroblasts proliferate and move into the wound, fibronectin in the upper granulation tissue assumes a more fine, linear reticular pattern disposed in a parallel array to the epidermis and to the fibroblasts themselves (Fig. 8). This fine reticular fibronectin is believed to emanate from fibroblasts, because of its close correlation in time of appearance and its intimate relationship with these cells (Fig. 8).

Fibroblasts produce an exuberant matrix in vitro, which looks similar to granulation tissue fibronectin, and presumably can do so in vivo under the right set of circumstances. Nevertheless, plasma fibronectin [180, 214] and fibrinogen-fibrin [215] can be incorporated into fibroblast matrix in vitro and such a process may occur in vivo. Indeed, plasma fibronectin does accumulate in normal extracellular matrices [216]. Furthermore, parallel arrays of fine reticular fibrin transiently occur coincident with the fine reticular fibronectin; however, this fibrin is largely removed before the fibronectin reaches maximal intensity [11]. In fact, Clark et al. [131] have shown that, although the bulk of fine reticular fibronectin in late cellular granulation tissue (7–10 days after injury) comes from in situ production, some matrix fibronectin even at this stage of wound healing derives from the plasma.

The factors that regulate the production of fibronectin by wound fibroblasts are unknown. Although thrombin can stimulate fibroblast proliferation and their synthesis and release of fibronectin [81, 82], it is not known whether thrombin continues to be generated in wounds after the initial fibrin clot formation. An additional possible candidate is epidermal growth factor (EGF), a polypeptide hormone concentrated in the salivary glands of male mice, which promotes fibronectin synthesis in vitro [217]. A second polypeptide hormone from salivary glands, nerve growth factor (NGF), has plasminogen activator activity and promotes wound contraction by an unknown mechanism [218]. Since fibronectin may form an important link between fibroblasts and matrix to facilitate wound contraction, NGF may enhance wound con-

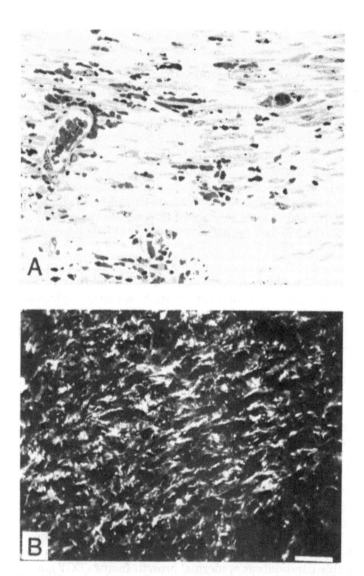

Figure 8 Granulation tissue of an excisional skin wound in guinea pigs at 7 days. The epidermis (not shown) is toward the top. (A) The 1 μm Giemsa-stained section illustrates the vertically oriented, dilated, newly formed blood vessels and the horizontally oriented, spindle-shaped, plump fibroblasts. Other cells in the tissue include macrophages, a few neutrophils, and erythrocytes. (B) Immunofluorescence micrograph of a cryostat section stained for fibronectin. The abundant fine fibronectin fibrils in the matrix are oriented largely parallel to the fibroblasts. Bars are 50 μm. (From Ref. 11.)

traction by the modulation of fibronectin synthesis or expression. Perhaps these two salivary polypeptides are the teleological reason for "licking one's wounds." Indeed, sialadenectomized mice heal skin wounds more slowly than controls, unless caged with normal mice [219]. Finally, exogenous fibronectin may itself stimulate fibroblasts to synthesize and secrete more endogenous fibronectin. Such a phenomenon is not without precedent, as fibronectin stimulates chondrocytes to produce more fibronectin[220, 221].

The molecular form of fibronectin in wounds is unknown. By microscopy, as noted, fibronectin of late cellular granulation tissue is in fine fibrils (Fig. 8B), corresponding to reticulin fibrils [9, 12], and decorates thicker collagen fibers in a periodic pattern [12]. Similar fibrils are found in vitro and consist chiefly of fibronectin, collagen and GAG. In analogy with culture matrix, one would predict that the wound matrix contains disulfide-linked fibronectin multimers [222, 223], factor XIII-catalyzed complexes, and noncovalent complexes with GAG, collagen, and other molecules, such as thrombospondin [224].

Several likely possibilities exist, none mutually exclusive, for the role of fibronectin in wound repair vis-a-vis the fibroblast. These possibilities include (1) enhancement of fibroblast adhesion to the matrix; (2) promotion of fibroblast migration by either chemotaxis or haptotaxis; (3) mediation of wound contraction; and (4) primary scaffolding for the deposition and orientation of collagen fibrils. The first three are probably different aspects of the same phenomenon. Since fibronectin is known to promote fibroblast adhesion in vitro to collagen (especially when denatured) and to fibrin, it no doubt does so in the wound matrix, initially in the fibrin gel and throughout areas of pre-existing denatured collagen, and later as a provisional matrix interlaced with newly synthesized types I and III collagen and hyaluronic acid.

Adhesion and migration are closely related, as demonstrated in experiments by Carter [225] with palladium gradients on glass. He found that cells migrate up an adhesive gradient on a substrate, a process he termed "haptotaxis" from the Greek word *heptein*, to grasp. One wonders whether the ability of fibronectin to promote directed fibroblast movement is not in fact chemotaxis, as presumed in the various reports [78, 78a, 226, 227], but haptotaxis. In fact, fibronectin enhancement of fibroblast movement was first noted on fibronectin-coated surfaces [184], and more recently fibroblasts have been observed to move along preformed fibronectin fibrils [228]. Thus, the mechanisms by which fibronectin stimulates directed fibroblast movement remain to be proved. Carter [225] also emphasized that too much adhesion would prevent migration. Indeed, such may be the case with fibroblasts and fibronectin according to Schor et al. [100], who found that fibronectin decreased the migration of human skin fibroblasts into a native type I collagen gel. Therefore, a high density of fibronectin and collagen might immobilize the fibroblast in the granulation tissue, but a gradient of submaximal levels would promote fibroblast migration by haptotaxis.

The most active form of fibronectin also deserves attention. In one study intact and cell binding fragments were about equally effective on a molar basis [229], whereas intact molecules seemed more effective in another [78]. A priori, one would expect the native soluble plasma form to be relatively inactive; otherwise directed movement away from vessels would be unlikely to occur [230]. The active form might be the insolubilized "solid-phase" fibronectin, a proteolytic fragment, or a new complex.

In recent years, wound contraction has been ascribed to specialized fibroblasts or myofibroblasts that contain actin microfilaments in the cytoplasm [231, 232]. These cells resemble fibroblasts in culture but are distinguished from normal tissue fibroblasts by their abundant bundles of actin and yet can be distinguished from smooth muscle cells by their abundant rough endoplasmic reticulum and Golgi apparati [233]. Such myofibroblasts are conspicuous in granulation tissue and are thought to mediate the visible musclelike contraction observed in isolated strips of granulation tissue exposed to serotonin (5-hydroxytryptamine), prostaglandin $F_{1\alpha}$, angiotensin, vasopressin, bradykinin, epinephrine, or norepinephrine and antagonized by prostaglandin E_1, papaverine, or cytochalasin B [233, 234]. Moreover, granulation tissue yields as much actinomyosin per gram as can be extracted from pregnant rat uterus [235].

The unified contraction of granulation tissue observed when isolated strips of tissue are exposed to pharmacological mediators theoretically necessitates molecular interconnections between the myofibroblasts. These may be either direct cell-cell connections or cell-substratum-cell links, or both. In fact, several types of cell-cell and cell-stroma linkages were observed by electron microscopy studies of granulation tissue from a 4-week-old open abdominal skin wound [233]. The cell-cell connections resembled desmosomes and gap junctions, whereas, the cell-substratum links appeared as extracellular microfilaments that ran between the cell and basal laminalike material that was frequently seen in close proximity to the cell surface and often in association with a dense zone in the microfilament bundle immediately beneath the plasma membrane. Ryan et al. stated, "The basal lamina material was not uncommonly seen running off into the stroma away from the cell (usually in a direction parallel to the intracytoplasmic microfilaments) and sometimes connecting with the basal lamina around another cell." They postulated that such "microtendon" structures participate in the transmission of cellular contraction to other tissue components and further proposed that the myofibroblasts contract as new collagen is laid down to give a "lock-step" system of tissue contraction [233]. Thus, the force of wound contraction is probably generated by actin bundles in the myofibroblasts and transmitted to the sides of the wound by cell-cell and cell-stroma links.

Majno later [231] pointed out the striking similarity between the micro-tendons of Ryan et al. and the fibronexus of Singer [236]. In a series of elegant transmission electron microscopic studies, Singer delineated the fibronexus as the colinear assemblage of intracytoplasmic 5 nm microfilaments and extracellular matrix fibrils in monolayers of human and hamster fibroblasts. Some of the chemical components comprising the fibronexus have been identified by immunoelectron microscopic investigations. Using indirect immunoferritin methods, fibronectin [236] and types I and III procollagen [237] were localized to the extracellular fibers of the fibronexus and actin [236] and vinculin [238] were identified as the major cytoplasmic constituents of the fibronexus. This work was actually in part a derivative from the double-label immunofluorescence experiments of Hynes and Destree [239] and Heggeness et al. [240], which demonstrated an extraordinary global coincidence and colinearity of actin and fibronectin in well-spread fibroblasts.

To determine whether the fibronexus is involved in fibroblast adhesion during wound healing, Singer and Clark have performed double labeled immunofluorescence microscopy with anti-actin and anti-fibronectin antibodies on 1 μm semithin frozen sections [241] and immunoelectron microscopy with anti-fibronectin antibodies on ultrathin sections [241] of granulation tissue formed in 7-9 day full-thickness, open guinea pig wounds. On immunofluorescence technique a large majority of actin fibers were observed to be coincident with fibronectin fibers. On immunoelectron microscopy these fibronectin fibers were localized around the surfaces of myofibroblasts and were prominent along attenuated myofibroblast processes that extended for some distance into the extracellular matrix of the granulation tissue. Three kinds of fibronexus or connections between myofibroblast intracytoplasmic actin filaments and extracellular fibronectin filaments occurred: (1) tandem associations between individual fibronectin fibers and microfilaments at the long ends of elongate myofibroblast processes, (2) plaquelike, and (3) tracklike fibronexus in which parallel fibronectin and actin fibers were connected by perpendicular transmembranous fibrils. From these data the authors concluded that the fibronexus may function as a major cohesive complex that serves to transmit the collective forces generated by the contraction of actin microfilaments within all the myofibroblasts of granulation tissue and thereby to affect wound contraction. In fact, the forces generated by wound contraction may greatly contribute to the extensive coalignment of actin and fibronectin fibers observed in granulation tissue.

A functional equivalent of wound contraction may exist in vitro in the form of collagen gels that slowly contract when fibroblasts are intermixed with the collagen just before gelation. The rate of contraction is proportional to the cell number and inversely proportional to the lattice collagen concentration [242]. When this phenomenon was studied by time-lapse photography it was found that the condensation of fibrils of the gel occurs as a result of a "collection" process executed by these cells as they extend and withdraw cytoplasmic podia that attach to collagen fibers. The latter are drawn toward the cell body in the course of podial contraction [243, 244]. Preliminary results show that fibronexuslike structures exist between these fibroblasts and the gel matrix (Clark, personal observations). Although exogenous fibronectin is not needed for this cell-mediated collagen contraction

(Clark, personal observations), a role for endogenous fibronectin has not been excluded. This system of in vitro gel contraction ultimately may add substantial information about how the force of contracting microfilaments of myofibroblasts is transmitted over a collagen matrix.

MATRIX REMODELING

The fibronectin matrix of early granulation tissue may serve as a template for collagen deposition as evidenced by the observations that fibroblasts tend to be aligned in the same axis as fibronectin (Fig. 8) [11, 12] and that collagen is later oriented in the same pattern. In addition, fibronectin in cell culture matrix is closely associated with collagen and GAG [245-248]. There are several lines of indirect evidence in vitro that collagen is deposited on a fibronectin matrix, not vice versa. McDonald et al. [249] noted that some fibrils in cultures stain for fibronectin but not collagen. Cells grown in the absence of ascorbate produce little or no collagen but form a dense fibronectin matrix [250]. Furthermore, collagenase releases collagen, but not fibronectin from fibroblast matrices [82], but thrombin releases both fibronectin and collagen. The release of collagen by thrombin is unlikely to be a direct effect on the collagen since thrombin does not digest native collagen but rather probably cleaves fibronectin or other glycoproteins that interact with collagen. In addition, an Fab' anti-fibronectin antibody, which blocks collagen binding to fibronectin, inhibited collagen accumulation in the extracellular matrix as judged by immunofluorescence and [3H]hydroxyproline incorporation [249]. However, fibronectin deposition was also inhibited, which indicates that the fibronectin and collagen interaction is necessary for fibril formation. Based on their data the authors suggest that "the deposition of fibrillar, extracellular fibronectin is essential for organization of types I and III collagen in vitro." If this applies in vivo, one might be able to influence the pattern of collagen formation by an artificial fibronectin template.

The in vivo interelationship of fibronectin and types I and III collagen deposition was reported in a study of cellulose sponge implants by Kurkinen et al. [6]. Sponge cubes 5 mm on a side were initially implanted in the peritoneal cavity of mice for 7 days and then transplanted subcutaneously into syngeneic mice. This "conditioning" promoted cellular and fibrous matrix penetration into the sponge. Such conditioned sponges initially contained an acidophilic protein-rich fluid that stained for fibronectin (but not collagen) in a needlelike pattern. Fibrin staining was not reported. Since a histological analog of the needlelike pattern was not seen in formalin-fixed material, the authors concluded it was an artifact. However, others have described, in similar sponge implantations, early formation of fibrils composed in part of fibrin that could be prevented by defibrination with ancrod [68, 69]. We suspect that the "conditioning" was caused by an accumulation of plasma-derived fibronectin-fibrin within the sponge that provided a better cellular substrate than the cellulose.

Fibroblasts invaded 1-2 mm into the sponge 7 days after subcutaneous transplantation and argyrophilic reticulin fibers that stained uniformly and strongly for fibronectin appeared concomitantly. Type III procollagen and some type I collagen was present but trailed behind the leading edge of fibroblasts. By 3-5 weeks, the fibroblasts had reached the center of the sponge, where their numbers were not greatest. In the central area the fibronectin and collagen distribution was as described in the sponge periphery at earlier times. However, by 3-5 weeks the now more mature periphery contained birefringent collagen bundles that stained for type I collagen. Fibronectin was diminished after 5 weeks as more mature birefringent collagen fibers were formed.

The term "granulation tissue" should probably not be used for this fibroplasia of sponge implants since capillaries, the hallmark of granulation tissue, are rare. Nonetheless, this important study documented the sequence in fibroplasia, now confirmed by others [7], of extracellular fibronectin deposition followed by cell invasion, more fibronectin, types I and III (recticulin) collagen, and finally type I (bifringent) collagen and loss of fibronectin. By immunoelectron microscopy the fibronectin in wounds is associated with fine fibrils [12] believed to be type III collagen and early forms of type I collagen that together form reticulin fibers [251].

The sequence of fibronectin followed by interstitial collagen has also been observed in wounds of children [18], in embryogenesis, and in certain pathological processes. In each, fibronectin

appears to be an early component, followed by type III collagen
and then type I collagen [6, 158, 252-254]. Fibronectin is often
associated with type III collagen in normal adult tissues, as in
blood vessels, papillary dermis, and lamina propria [158]. Two
notable exceptions are the hepatic sinusoids and glomerular
mesangium: both have fibronectin with little or no detectable type
III collagen [255]. The significance of the discordance is unclear.
In pathological conditions of increased fibrosis, such as hypertrophic scars, keloids, scleroderma, or liver cirrhosis, type III collagen and fibronectin are conjointly increased [251, 256-258]. As
proposed by Holund et al. [7] and others, the presence of fibronectin may indicate an early stage of tissue repair or ongoing fibroplasia.

The association of fibronectin and type III collagen in vivo
correlates with the higher avidity in vitro of fibronectin for native
type III collagen, among the various native collagen types. The
greatest preference is for denatured type III. The order of binding
avidity is dIII > dI, dII, dIV, nIII > dV > nI, nII, nIV (n and d indicate native or denatured collagen type, respectively) [259, 260].
Direct translation into in vivo conditions is hazardous, because the
"native" forms used in vitro are chemically isolated and other
molecules, particularly GAG, can markedly increase the binding
between fibronectin and collagen [260-262].

OTHER CELLS

Neutrophils

Neutrophils participate in wounds primarily as phagocytic and an-
tibacterial cells [97] and may utilize fibronectin to ingest bacteria
[263]. The first step, however, in the accumulation of leukocytes
at a site of inflammation or injury, is cell adhesion to the blood
vessel endothelium at that site. The role of fibronectin if any, in
leukocyte adhesion to endothelial cells has been investigated in
several laboratories. Fibronectin, treated with dithiothreitol and
allowed to aggregate, promotes adherence of neutrophils to cultured endothelial cells [264], but untreated fibronectin either has
no effect (Tonnesen and Clark, personal observations) [264] or
inhibits sticking [265]. These variations in neutrophil binding to

fibronectin [266] seem to reflect differences in the molecular form of fibronectin and may reflect changes in the state of the neutrophil membrane due to isolation procedures.

Mast Cells

Although the normal biological role of mast cells remains enigmatic, they do increase in numbers during wound healing, tumor growth, and chronic inflammation. Mast cell accumulation provides increased stores of histamine, arachidonic acid metabolites, heparin, and proteases to these tissue sites. Mast cell chymase and cathepsin G have been demonstrated to degrade both soluble and matrix fibronectin [267]. These mast cell proteases may cleave pericellular matrices in vivo, thus freeing cells to migrate to other sites. In addition, the glycosaminoglycan heparin binds to fibronectin at two molecular domains, and this complex can form a fibrillar precipitate. That such complexes may occur in vivo on the mast cell membrane is suggested by the finding that rat peritoneal mast cells have fibronectin on their surface after gradient isolation as determined by immunofluorescence and induction of histamine release by anti-fibronectin antibodies [268, 269]. Whether this is restricted to peritoneal mast cells and whether fibronectin is functionally important are unknown. Evidence has been advanced that heparin is an important cofactor in angiogenesis and is supported by the fact that high doses of protamine to antagonize neovascularization in experimental tumors [195]. Protamine also precipitates fibronectin, and alterations of fibronectin may be involved in these effects.

Neurite Outgrowth

Neurite outgrowth [270, 271] and Schwann cell [272, 273] and neural crest cell migration [274] in culture are promoted by substrate fibronectin, and it is likely that fibronectin plays a similar contact guidance role in wounds. The observation that plasma clots could be used as "sutures" to increase axon migration across transected nerves [275] is being re-examined by Matras [276] and others in the light of this new basic knowledge.

Although knowledge of fibronectin in wounds involving bones and other organs is still limited, analogous fibronectin roles are expected. One additional important effect, potentially relevant to these wounds, is the ability of fibronectin to inhibit differentiation of mesenchymal cells. Fibronectin added to cultured chondroblasts [220, 221] or myoblasts [277] prevents synthesis of type II collagen or fusion into myotubes, respectively. These effects are most relevant to limb regeneration in amphibians [15], but a case can be made for a dedifferentiation effect of epithelial cells, fibroblasts, and endothelial cells in mammalian wounds.

DEFECTIVE WOUND HEALING

Analysis of hereditary molecular deficiences can offer decisive proof of the in vivo role of individual molecules. In the case of fibronectin, no absolute hereditary deficiency of plasma or cellular fibronectin levels has been described. In all likelihood, such absolute deficiency is incompatible with viable embryonic differentiation. However, a kindred, who inherits hyperextensible joints and skin and defective wound healing in an autosomal recessive pattern, has an associated abnormal plasma fibronectin [278]. A brief discussion of this important observation is warranted.

The propositus, a 28-year-old woman, has hyperextensible joints, mitral valve prolapse, and easy bruisability but normal menstruation and hemostasis. She has thin hyperextensible skin and a defect in wound healing characterized by wound separation ("fishmouth scarring"). This patient was classified as having a new type of Ehlers-Danlos syndrome (type IX). Her platelets and those from her three affected siblings showed little or no aggregation in response to collagen or ADP, but no clear abnormality in response to thrombin or epinephrine. Plasma fibronectin antigen levels were normal; however, the defect in collagen-induced platelet aggregation could be restored by addition of purified plasma fibronectin from another individual, suggesting that the fibronectin molecule itself is abnormal. Nevertheless, the necessity of exogenous fibronectin for platelet aggregation is debated (see Chap. 8).

Preliminary characterization of the affected kindred's plasma fibronectin suggests that the amino-terminal heparin binding domain is less basic (Furcht, personal communication). The nature of this alteration (carbohydrate, amino acid sequence, or attached molecule) has not been established. It would be of obvious interest

to analyze the fibronectin further and determine whether exogenous plasma fibronectin could improve wound healing in this kindred.
    A second hereditary defect, factor XIII deficiency, has also
been associated with poor wound healing [86]. As discussed, the
mechanism may involve failure to cross-link fibronectin to fibrin
or collagen in wounds [54]. Unfortunately, little more is known
about the wound morphology and fibronectin localization in such
patients.
    In the future, numerous other abnormalities in wound healing,
either acquired or hereditary, such as keloids and hypertrophic
scars, chronic corneal ulceration, and diabetic ulcers, will probably
be examined for possible abnormalities in fibronectin.

POSSIBLE PROMOTION OF WOUND HEALING BY EXOGENOUS FIBRONECTIN

The observations of fibronectin in wounds have led to attempts to
promote wound healing by local addition of exogenous fibronectin. Weiss et al. [279] have reported that exogenous plasma fibronectin applied as eye drops caused a modest but significant acceleration of epithelial migration in rabbit corneal ulceration. Because of the strong evolutionary pressure on the ability to heal
efficiently, few if any means are known that will accelerate healing
of sterile, well-apposed, uncomplicated wounds. Thus, it is not unexpected that exogenous fibronectin would have only a small effect on simple corneal scrape wounds, since abundant endogenous
fibronectin is deposited normally. Of greater interest and potential
therapeutic importance would be to study conditions of defective
wound healing. Clinical trials of local fibronectin administation in
severe corneal ulceration have begun in Japan. In vitro studies by
Nishida et al. [280] have suggested that exogenous plasma fibronectin promotes migration of corneal epithelial cells over the cut
surface of the corneal stroma. Other studies, however, have been
unable to observe any effect of exogenous fibronectin on in vitro
corneal scrape wounds (Fujikawa et al., unpublished).
    Nevertheless, the experimental approach of applying a local
preformed fibrin-fibronectin gel described seems scientifically
sound and clinically promising [276]. Systemic administration of
fibronectin is less likely to have important effects on wound heal-

ing, since infants, who have more rapid wound healing than adults, have about 35% adult levels of plasma fibronectin [281]. To plan and interpret future experiments, certain critical information will be needed. First, the actual form(s) of fibronectin in wounds should be determined by direct analysis and the biological activity of such form(s) demonstrated in vitro. Furthermore, in therapeutic trials it will be important to document the status of both the endogenous fibronectin and the incorporation of the exogenous fibronectin in the wound site.

CONTROL MECHANISMS

Despite, or because of, the numerous interactions of fibronectin demonstrable in vitro, the roles of fibronectin in vivo are very complex [Fig. 9]. It is puzzling that the diverse cellular responses to fibronectin, such as attachment, chemotaxis, aggregation, differentiation, and phagocytosis, would be stimulated by native, circulating fibronectin, without requiring a modification to an active form; otherwise such processes would occur continuously in vivo. For example, how is the chemotactic activity of fibronectin for endothelial cells and fibroblasts manifested in vivo, when fibronectin circulates in the plasma at levels higher than needed for chemotaxis in vitro? Although definite answers are not at present possible, several studies suggest that alteration in the molecular form of the fibronectin is a critical determinant of its biological activity. Analogous control mechanisms can be found for many other biological molecules, including chemotactic factors and opsonins.

As described, fibronectin exists in many well-recognized forms, including multimers, dimers, fragments, and complexes with other molecules, and important conformation changes may distinguish fibronectin in liquid and solid phase. Elsewhere we have presented the limited, but provocative, evidence for the hypothesis that fibronectin may be converted to more active forms and that this may determine the biological effects [230]. For example, fibronectin *fragments* are more active in monocyte chemotaxis [76] and promotion of opsonization of alternative complement pathway activators [282]. Macrophages [107, 108, 117] and fibroblasts [28, 283] bind preferentially to *solid-phase* fibronectin; that is, soluble fibronectin does not block the attachment of

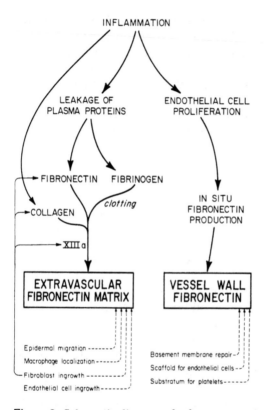

Figure 9 Schematic diagram of a few proposed roles for fibronectin in inflammatory processes, including wound healing, based on evidence and hypotheses given in the text. (From Ref. 25.)

these cells to fibronectin-coated substrates. The increased activity
of fibronectin bound to a solid phase may be explained by multivalent interactions or conformational changes of the fibronectin.
Analogous mechanisms are thought to release latent activity of
other plasma proteins, such as complement activation by aggregated IgG (multivalent binding) [284] and activation of factor XII
by attachment to a solid phase (conformation change) [285, 286].
Careful testing of this hypothesis will require analysis of the forms
of fibronectin in vivo and further demonstration that the form in-

fluences activity in vitro. Such an approach should shed light on the regulation of the potential functions of fibronectin and provide a more rational basis of therapeutic intervention in those biological processes that involve fibronectin.

REFERENCES

1. E. Spiegel, M. Burger, and M. Spiegel, Fibronectin in the developing sea urchin embryo. *J. Cell Biol. 87*:309-313 (1980).
2. J. Wartiovaara, I. Leivo, and A. Vaheri, Expression of the cell surface-associated glycoprotein, fibronectin, in the early mouse embryo. *Dev. Biol. 69*:247-257 (1979).
3. E. Linder, A. Vaheri, E. Ruoslahti, and J. Wartiovaara, Distribution of fibroblast surface antigen in the developing chick embryo. *J. Exp. Med. 142*:41-49 (1975).
4. W. T. Gibson, J. R. Couchman, R. A. Badley, H. J. Saunders, and C. G. Smith, Fibronectin in cultured rat keratinocytes: Distribution, synthesis and relationship to cytoskeletal proteins. *Eur. J. Cell. Biol. 30*:205-213 (1983).
5. M. G. Tonnesen, S. L. Siegal, L. A. Lee, J. C. Huff, and R. A. F. Clark, Expression of fibronectin and factor VIII related antigen in the human cutaneous microvasculature changes during development. *Clin. Res. 32*: 143A (1984).
6. M. Kurkinen, A. Vaheri, P. J. Roberts, and S. Steinman, Sequential appearance of fibronectin and collagen in experimental granulation tissue. *Lab. Invest. 43*:47-51 (1980).
7. B. Holund, I. Clemmensen, P. Junker, and H. Lyon, Fibronectin in experimental granulation tissue. *Acta Pathol. Microbiol. Immunol. Scand. 90*:159-165 (1982).
8. R. A. F. Clark, J. H. Quinn, H. J. Winn, J. M. Lanigan, p. DellePella, and R. B. Colvin, Fibronectin is produced by blood vessels in response to injury. *J. Exp. Med. 156*:646-651 (1982).
9. F. Grinnell, R. E. Billingham, and L. Burgess, Distribution of fibronectin during wound healing in vivo. *J. Invest. Dermatol. 76*:181-189 (1981).
10. R. A. F. Clark, J. M. Lanigan, P. DellePella, E. Manseau, H. F. Dvorak, and R. B. Colvin, Fibronectin and fibrin provide a provisional matrix for epidermal cell migration during wound reepithelialization. *J. Invest. Dermatol. 79*:264-269 (1982).
11. R. A. F. Clark, P. DellePella, E. Manseau, J. M. Lanigan, H. F. Dvorak, and R. B. Colvin, Blood vessel fibronectin increases in conjunction with

endothelial cell proliferation and capillary ingrowth during wound healing. *J. Invest. Dermatol.* 79:269–276 (1982).

12. L. A. Repesh, T. J. Fitzgerald, and L. T. Furcht, Fibronectin involvement
in granulation tissue and wound healing in rabbits. *J. Histochem. Cytochem.* 30:351–358 (1982).

13. L. S. Fujikawa, C. S. Foster, T. J. Harrist, J. M. Lanigan, and R. B. Colvin, Fibronectin in the healing rabbit corneal wound. *Lab. Invest.* 45:
120–129 (1981).

14. L. S. Fujikawa, C. S. Foster, I. K. Gipson, and R. B. Colvin, Basement
membrane components in healing rabbit corneal epithelial wounds: Immunofluorescence and ultrastructural studies. *J. Cell Biol.* 98:128–
138 (1984).

15. L. A. Repesh, T. J. Fitzgerald, and L. T. Furcht, Changes in the distribution of fibronectin during limb regeneration in newts using immunochemistry. *Differentiation* 22:125–131 (1982).

16. A. K. Gulati, A. A. Zalewski, and A. H. Reddi, An immunofluorescent
study of the distribution of fibronectin and laminin during limb regeneration in the adult newt. *Dev. Biol.* 96:355–365 (1983).

17. D. J. Donaldson and J. T. Mahan, Fibrinogen and fibronectin as substrates for epidermal cell migration during wound closure. *J. Cell Sci.* 62:
117–127 (1983).

18. J. Viljanto, R. Penttinen, and J. Raekallio, Fibronectin in early phases of
wound healing in children. *Acta Chir. Scand.* 147:7–13 (1981).

19. J. A. Schilling, Wound healing. *Surg. Clin. North Am.* 56:859–874
(1976).

20. E. Peacock and W. van Winkle, *Surgery and Biology of Wound Repair.*
W. B. Saunders, Philadelphia, 1976.

21. H. R. Hunt, *Wound Healing and Wound Infection: Theory and Surgical
Practice.* Appleton-Century-Crofts, New York, 1980.

22. L. E. Glynn (ed.), *Tissue Repair and Regeneration, Handbook of Inflammation,* Vol. III. Elsevier/North Holland Biomedical Press, Amsterdam,
1981.

23. S. Shoshan, Wound healing. *Int. Rev. Connect. Tiss Res.* 9:1–26 (1981).

24. F. Grinnell, Fibronectin and wound healing. *Am. J. Dermatopathol.* 4:
185–188 (1982).

25. R. B. Colvin, Roles of fibronectin in wound healing, in *Fibronectin*
(D. F. Mosher, ed.). Academic Press, New York.

26. K. Yamada and K. Olden, Fibronectins-adhesive glycoproteins of cell surface and blood. *Nature* 275:179–184 (1978).

27. R. J. Klebe, Isolation of a collagen-dependent cell attachment factor.
Nature 250:248–251 (1974).

28. E. Pearlstein, Plasma membrane glycoprotein which mediates adhesion
of fibroblasts to collagen. *Nature* 262:497–500 (1976).

245

29. F. Grinnell and M. K. Feld, Initial adhesion of human fibroblasts in serum-free medium: Possible role of secreted fibronectin. *Cell 17*:117–129 (1979).
30. K. Sekiguchi, M. Fukuda, and S. Hakomori, Domain structure of hamster plasma fibronectin. Isolation and characterization of four functionally distinct domains and their unequal distribution between two subunit polypeptides. *J. Biol. Chem. 256*:6452–6462 (1981).
31. H. Hörmann and M. Seidl, Affinity chromatography on immobilized fibrin monomer. III. The fibrin affinity ceter of fibronectin. *Hoppe-Seylers Z. Physiol. Chem. 361*:1449–1452 (1980).
32. M. Hayashi and K. M. Yamada, Domain structure of the carboxyl-terminal half of human plasma fibronectin. *J. Biol. Chem. 258*:3332–3340 (1983).
33. K. Sekiguchi and S. I. Hakomori, Domain structure of plasma fibronectin. *J. Biol. Chem. 258*:3967–3973 (1983).
34. E. Ruoslahti, E. G. Hayman, P. Kuusela, J. E. Shively, and E. Engvall, Isolation of a tryptic fragment containing the collagen-binding site of plasma fibronectin. *J. Biol. Chem. 254*:6054–6059 (1979).
35. G. Balian, E. M. Click, E. Crouch, J. M. Davidson, and P. Bornstein, Isolation of a collagen-binding fragment from fibronectin and cold-insoluble globulin. *J. Biol. Chem. 254*:1429–1432 (1979).
36. L. H. E. Hahn and K. M. Yamada, Identification and isolation of a collagen-binding fragment of the adhesive glycoprotein-fibronectin. *Proc. Natl. Acad. Sci. USA 76*:1160–1163 (1979).
37. L. I. Gold, A. Garcia-Pardo, B. Frangione, E. C. Franklin, and E. Pearlstein, Subtilisin and cyanogen bromide cleavage products of fibronectin that retain gelatin-binding activity. *Proc. Natl. Acad. Sci. USA 76*:4803–4807 (1979).
38. M. B. Furie, A. B. Frey, and D. B. Rifkin, Location of a gelatin-binding region of human plasma fibronectin. *J. Biol. Chem. 255*:4391–4394 (1980).
39. M. D. Pierschbacher, E. G. Hayman, and E. Ruoslahti, Location of the cell-attachment site in fibronectin with monoclonal antibodies and proteolytic fragments of the molecule. *Cell 26*:259–267 (1981).
40. M. D.Pierschbacher, E. Ruoslahti, J. Sundelin, P. Lind, and P. A. Peterson, The cell attachment domain of fibronectin. Determination of the primary structure. *J. Biol. Chem. 257*:9593–9597 (1982).
41. L. Kjellen, I. Petterson, and M. Hook, Cell-surface heparan sulfate: An intercalated membrane proteoglycan. *Proc. Natl. Acad. Sci. USA 78*:5371–5375 (1981).
42. J. Laterra, R. Ansbacher, and L. A. Culp, Glycosaminoglycans that bind cold-insoluble globulin in cell-substratum adhesion sites of murine fibroblasts. *Proc. Natl. Acad. Sci. USA 77*:6662–6666 (1980).

43. R. J. Klebe and P. J. Mock, Effect of glycosaminoglycans on fibronectin-mediated cell attachment. *J. Cell Physiol. 112*:5-9 (1982).

44. M. Hayashi and K. M. Yamada, Divalent cation modulation of fibronectin binding to heparin and to DNA. *J. Biol. Chem. 257*:5263-5267 (1982).

45. H. Richter, M. Seidl, and H. Hormann, Location of heparin-binding sites of fibronectin. Detection of a hitherto unrecognized transamidase sensitive site. *Hoppe-Seylers Z. Physiol. Chem. 362*:399-408 (1981).

46. F. Jilek and H. Hormann, Fibronectin (cold-insoluble globulin), influence of heparin and hyaluronic acid on binding of native collagen. *Hoppe-Seylers Z. Physiol. Chem. 360*:597-603 (1979).

47. K. M. Yamada, D. W. Kennedy, K. Kimata, and R. M. Pratt, Characterization of fibronectin interactions with glycosaminoglycans and identification of active proteolytic fragments. *J. Biol. Chem. 255*:6055-6063 (1980).

48. H. Hörmann, and V. Jelinic, Regulation by heparin and hyaluronic acid of the fibronectin-dependent association of collagen, Type III, with macrophages. *Hoppe-Seylers Z. Physiol. Chem. 362*:87-94 (1981).

49. B. Voss, S. Allam, S., Rauterberg, J., Ullrich, K., Gieselmann, V., and von Figura, K. Primary cultures of rat hepatocytes synthesize fibronectin. *Biochem. Biophys. Res. Commun. 90*:1348-1354 (1979).

50. J. W. Tamkin and R. O. Hynes, Plasma fibronectin is synthesized and secreted by hepatocytes. *J. Biol. Chem. 258*:4641-4647 (1983).

51. M. R. Owens and C. D. Cimino, Synthesis of fibronectin by the isolated perfused rat liver. *Blood 59*, 1305-1309 1982.

52. J. Laterra and L. A. Culp, Differences in hyaluronate binding to plasma and cell surface fibronectins. Requirement for aggregation. *J. Biol. Chem. 257*:719-726 (1982).

53. D. F. Mosher, Action of fibrin-stabilizing factor on cold-insoluble globulin and α_2-macroglobulin in clotting plasma. *J. Biol. Chem. 251*, 1639-1645 (1976).

54. D. F. Mosher, P. E. Schad, and H. K. Kleinman, Cross-linking of fibronectin to collagen by blood coagulation factor XIIIa. *J. Clin. Invest. 64*: 781-787 (1979).

55. D. F. Mosher, Cross-linking of cold-insoluble globulin by fibrin-stabilizing factor. *J. Biol. Chem. 250*:6614-6621 (1975).

56. F. Grinnell, M. Feld, and D. Minter, Fibroblast adhesion to fibrinogen and fibrin substrata: Requirement for cold-insoluble globulin (plasma fibronectin). *Cell 19*:517-525 (1980).

57. F. Duckert, Documentation of the plasma factor XIII deficiency in man. *Ann. N. Y. Acad. Sci. 202*:190-199 (1972).

58. D. D. Wagner and R. O. Hynes, Domain structure of fibronectin and its relation to function. *J. Biol. Chem. 254*:6746-6754 (1979).

59. J. A. McDonald and D. G. Kelley, Degradation of fibronectin by human leukocyte elastase. Release of biologically active fragments. *J. Biol. Chem.* 255:8848-8858 (1980).

60. D. D. Wagner and R. O. Hynes, Topological arrangement of the major structural features of fibronectin. *J. Biol. Chem.* 255:4304-4312 (1980).

61. N. M. Tooney, M. W. Mosesson, D. L. Amrani, J. F. Hainfeld, and J. S. Wall, Solution and surface effects on plasma fibronectin structure. *J. Cell Biol.* 97:1686-1692 (1983).

62. H. P. Erickson and N. A. Carrell, Fibronectin in extended and compact conformations. Electron microscopy and sedimentation analysis. *J. Biol. Chem.* 258:14539-14544 (1983).

63. M. Rocco, M. Carson, R. Hantgan, J. McDonagh, and J. Hermans, Dependence of the shape of the plasma fibronectin molecule on solvent composition. *J. Biol. Chem.* 258:14545-14549 (1983).

64. R. A. F. Clark, H. F. Dvorak, and R. B. Colvin, Fibronectin in delayed-type hypersensitivity skin reactions: Associations with vessel permeability and endothelial cell activation. *J. Immunol.* 126:787-793 (1981).

65. W. R. Bell, Defibrinogenating enzymes, in *Hemostasis and Thrombosis, Basic Principles and Clinical Practice* (R. W. Colman, J. Hirsch, V. J. Marder, and E. W. Salzman, eds.). J. B. Lippincott, Philadelphia, 1982, pp. 1013-1027.

66. L. A. Sherman and J. Lee, Fibronectin: Blood turnover in normal animals and during intravascular coagulation. *Blood* 60:558-563 (1982).

67. T. M. Daniel, S. V. Pizzo, and P. A. McKee, Comparison of ancrod and heparin as anticoagulants following endarterectomy in the dog. *Ann. Surg.* 184:223-228 (1976).

68. S. Brandstedt and P. S. Olson, Effect of defibrinogenation on wound strength and collagen formation. A study in the rabbit. *Acta. Chir. Scand.* 146:483-486 (1980).

69. S. Brandstedt, F. Rank, and P. S. Olson, Wound healing and formation of granulation tissue in normal and defibrinogenated rabbits. An experimental model and histological study. *Eur. Surg. Res.* 12:12-21 (1980).

70. S. Brandstedt and P. S. Olson, Lack of influence on collagen accumulation in granulation tissue with "delayed" defibrinogenation. A study in the rabbit. *Acta Chir. Scand.* 147:89-91 (1981).

71. J. E. Kaplan, J. Molnar, T. M. Saba, and C. Allen, Comparative disappearance and localization of isotopically labelled opsonic protein and soluble albumin following surgical trauma. *J. Reticuloendothel. Soc.* 20:375-384 (1976).

72. A. B. Robbins, J. E. Doran, A. C. Reese, and A. R. Mansburger, Jr. Cold insoluble globulin levels in operative trauma: Serum depletion, wound sequestration, and biological activity: An experimental and clinical study. *Ann. Surg.* 46:663-672 (1980).

73. R. A. F. Clark, H. F. Dvorak, M. Mosesson, and R. B. Colvin, Fibronectin deposition in delayed-type hypersensitivity reactions of normals and a patient with afibrinogenemia. *J. Clin. Invest. 74*:1011-1016 (1984).

74. D. Menache, Abnormal fibrinogens. A review. *Thromb. Diath. Haemorrh. 29*:525-535 (1973).

75. S. K. Banergee and L. E. Glynn, Reactions to homologous and heterologous fibrin implants in experimental animals. *Ann. N. Y. Acad. Sci. 86*: 1064-1074 (1980).

76. D. A. Norris, R. A. F. Clark, L. M. Swigart, J. C. Huff, W. L. Weston, and S. E. Howell, Fibronectin fragment(s) are chemotactic for human peripheral blood monocytes. *J. Immunol. 129*:1612-1618 (1982).

77. J. C. Bowersox and N. Sorgente, Chemotaxis of aortic endothelial cells in response to fibronectin. *Cancer Res. 42*:2547-2551 (1982).

78. A. E. Postlethwaite, J. Keski-Oja, G. Balian, and A. H. Kang, Induction of fibroblast chemotaxis by fibronectin. Localization of the chemotactic region to a 140,000-molecular weight non-gelatin-binding fragment. *J. Exp. Med. 15*:494-499 (1981).

78a. H. E. J. Seppa, K. M. Yamada, S. T. Seppa, M. H. Silver, H. K. Kleinman, and E. Schiffman, The cell binding fragment of fibronectin in chemotactic for fibroblasts. *Cell Biol. Int. Rep. 5*:813-819 (1981).

79. A. B. Kay, D. S. Pepper, and R. McKenzie, The identification of fibrinopeptide B as a chemotactic agent derived from human fibrinogen. *Br. J. Haematol. 27*:669-677 (1974).

80. R. Bar-Shavit, A. Kahn, J. W. Fenton, and G. D. Wilner, Chemotactic response of monocytes to thrombin. *J. Cell Biol. 96*:282-285 (1983).

81. D. F. Mosher and A. Vaheri, Thrombin stimulates the production and release of a major surface-associated glycoprotein (fibronectin) in cultures of human fibroblasts. *Exp. Cell Res. 112*:323-334 (1978).

82. J. Keski-Oja, G. J. Todaro, and A. Vaheri, Thrombin affects fibronectin and procollagen in the pericellular matrix of cultured human fibroblasts. *Biochim. Biophys. Acta 673*:323-331 (1981).

83. P. B. Bitterman, S. I. Rennard, S. Adelberg, and R. G. Crystal, Role of fibronectin as a growth factor for fibroblasts. *J. Cell Biol. 97*:1925-1932 (1983).

84. H. Seppa, G. Grotendorst, S. Seppa, E. Schiffman, and G. R. Martin, Platelet-derived growth factor is chemotactic for fibroblasts. *J. Cell Biol. 92*:584-588 (1982).

85. T. F. Deuel, R. M. Senior, D. Chang, G. L. Griffin, R. L. Heinrikson, and E. T. Kaiser, Platelet factor 4 is chemotactic for neutrophils and monocytes. *Proc. Natl. Acad. Sci. 78*:4584-4587 (1981).

86. L. Lorand, M. S. Losowsky, and K. J. M. Miloszewski, Human factor XIII: Fibrin-stabilizing factor. *Prog. Hemost. Thromb. 5*:245-290 (1980).

87. J. Sakata and N. Aoki, Cross-linking of alpha 2-plasmin inhibitor to fibrin by fibrin-stabilizing factor. *J. Clin. Immunol.* 65:290-297 (1980).

88. D. R. Kahn and I. Cohen, Factor XIII a-catalyzed coupling of structural proteins. *BBA* 668:490-494 (1981).

89. E. Beck, F. Duckert, and M. Ernst, The influence of fibrin stabilizing factor on the growth of fibroblasts in vitro and wound healing. *Thromb. Diathes. Haemorrh,* 6:485-492 (1961).

90. H. D. Bruhn and Polh, Growth regulation of fibroblasts by thrombin, factor XIII and fibronectin. *Klin. Wochenschr.* 59:145-146 (1981).

91. G. Schroff, C. Neuman, and C. Sorg, Transglutaminase as a marker for subsets of murine macrophages. *Eur. J. Immunol.* 11:637-642 (1981).

92. P. J. Birckbichler and M. K. Patterson, Cellular transglutaminase, growth, and transformation. *Ann. N. Y. Acad. Sci.* 312:354-365 (1978).

93. J. E. Folk and J. S. Finlayson, The ε-(γ-glutamyl)lysine crosslink and the catalytic role of transglutaminases. *Adv. Protein Chem.* 31:1-133 (1977).

94. V. H. Biel, H. Bohn, H. Ronneberger, and O. Zwisler, Beschleunigung der wundheilung durch faktor XIII der Blutgerinnung. *Arznein. Forsch.* 21: 1429-1430 (1971).

95. W. Marktl and B. Rudas, The effect of factor XIII on wound granulation in the rat. *Thromb. Diath. Haemorrh.* 32:578-581 (1974).

96. V. H. Knoche and G. Schmidt, Autoradiographische untersuchungen uber den einfluss des faktors XIII auf die Wundheilung im Tierexperiment. *Arzneim. Forsch.* 26:547-551 (1976).

97. S. J. Liebovich and R. Ross, The role of the macrophage in wound repair. A study with hydrocortisone and antimacrophage serum. *Am. J. Pathol.* 78:71-100 (1975).

98. B. M. Martin, M. A. Gimbrone, Jr., E. R. Unanue, and R. S. Cotran, Stimulation of nonlymphoid mesenchymal cell proliferation by a macrophage-derived growth factor. *J. Immunol.* 126:1510-1515 (1981).

99. B. M. Martin, M. A. Gimbrone, G. R. Majeau, E. R. Unanue, and R. S. Cotran, Stimulation of human monocyte/macrophage-derived growth factor (MDGF) production by plasma fibronectin. *Am. J. Patho.* 111: 367-373 (1983).

100. S. L. Schor, A. M. Schor, and G. W. Bazill, The effects of fibronectin on the migration of human foreskin fibroblasts and Syrian hamster melanoma cells into three-dimensional gels of native collagen fibres. *J. Cell Sci.* 48:301-314 (1981).

101. J. A. Schmidt, S. B. Mizel, D. Cohen, and I. Green, Interleukin 1, a potential regulator of fibroblast proliferation. *J. Immunol.* 128:2177-2182 (1982).

102. A. E. Postlethwaite, L. B. Lachman, C. L. Mainardi, and A. H. Kang, Interleukin 1 stimulation of collagenese production by cultured fibroblasts. *J. Exp. Med.* 157:801-806 (1983).

103. R. B. Colvin, J. M. Lanigan, R. A. F. Clark, T. H. Ebert, E. Verderber, and M. E. Hammond, Macrophage fibronectin (cold-insoluble globulin, LETS protein). *Fed. Proc. 38*:1408 (1979).
104. S. Johansson, K. Rubin, M. Hook, T. Ahlgren, and R. Seljelid, In vitro biosynthesis of cold insoluble globulin (fibronectin) by mouse peritoneal macrophages. *FEBS Lett. 105*:313-316 (1979).
105. K. Alitalo, T. Hovi, and A. Vaheri, Fibronectin is produced by human macrophages. *J. Exp. Med. 151*:602-613 (1980).
106. A. E. Postlethwaite, J. M. Seyer, and A. H. Kang, Chemotactic attraction of human fibroblasts to type I, II and III collagens and collagen-derived peptides. *Proc. Natl. Acad. Sci. USA 75*:871-875 (1978).
107. M. P. Bevilacqua, D. Amrani, M. W. Mosesson, and C. Bianco, Receptors for cold-insoluble globulin (plasma fibronectin) on human monocytes. *J. Exp. Med. 153*:42-60 (1981).
108. C. R. Horsburg, C. H. Kirkpatrick, and R. A. F. Clark, Platelets modulate human macrophage adhesiveness. *Clin. Res. 30*:318A (1982).
109. P. W. Gudewicz, J. Molnar, M. Z. Lai, D. W. Beezhold, G. E. Siefrihg, R. B. Crede, and L. Lorand, Fibronectin-mediated uptake of gelatin-coated latex particles by peritoneal macrophages. *J. Cell Biol 87*:427-433 (1980).
110. F. A. Blumenstock, T. M. Saba, E. Roccario, E. Cho, and J. E. Kaplan, Opsonic fibronectin after trauma and particle injection determined by a peritoneal macrophage monolayer assay. *J. Reticuloendothel. Soc. 30*:61-71 (1981).
111. J. E. Doran, A. R. Mansberger, H. T. Edmondson, and A. C. Reese, Cold insoluble globulin and heparin interactions in phagocytosis by macrophage monolayers: Lack of heparin requirement. *J. Reticuloendothel. Soc. 29*:275-283 (1981).
112. L. Van de Water, S. Schoeder, E. B. Crenshaw, and R. O. Hynes, Phagocytosis of gelatin-latex particles by a murine macrophage line is dependent on fibronectin and heparin. *J. Cell. Biol. 90*:32-39 (1981).
113. B. Villiger, D. G. Kelley, W. Engleman, C. Kuhn, and J. A. McDonald, Human alveolar macrophage fibronectin: Synthesis, secretion, and ultrastructural localization during gelatin-coated latex particle binding. *J. Cell Biol. 90*:711-720 (1981).
114. H. P. Godfrey and A. Purohit, Reversible binding of a guinea-pig lymphokine to gelatin and fibrinogen: Possible relationship of macrophage agglutination factor and fibronectin. *Immunology 46*:507-514:515-526 (1982).
115. H. G. Remold, J. E. Shaw, and J. R. David, A macrophage surface component related to fibronectin is involved in the response to migration inhibitory factor. *Cell. Immunol. 58*:175-187 (1981).

116. R. B. Colvin and H. F. Dvorak, Roles of the clotting system in cell-mediated hypersensitivity. II. Kinetics of fibrinogen/fibrin accumulation and vascular permeability changes in tuberculin and cutaneous basophil hypersensitivity reactions. *J. Immunol. 114*:377-387 (1975).

117. R. B. Colvin, Fibrinogen-fibrin interactions with fibroblasts and macrophages. *Ann. N. Y. Acad. Sci. 408*:621-634 (1983).

118. D. Billig, A. Nicol, R. McGinty, P. Cowin, J. Morgan, and D. Garrod, The cytoskeleton and substratum adhesion in chick embryonic corneal epithelial cells. *J. Cell Sci. 57*:51-71 (1982).

119. K. A. Holbrook, Structure and function of the developing human skin, in *Biochemistry and Physiology of the Skin* (L. A. Goldsmith, ed.). Oxford University Press, New York, 1983 pp. 64-101.

120. K. Naito, S. Morioka, and H. Ogawa, The pathogenic mechanisms of blister formation in bullous pemphoid. *J. Invest. Dermatol. 79*:303-306 (1982).

121. M. Masutani, H. Ogawa, A. Taneda, M. Shoji, and H. Miyazaki, Ultrastructural localization of immunoglobulins in the dermo-epidermal junction of a patient with bullons pemphigoid. *J. Dermatol. 5*:107-112 (1978).

122. G. Schaumburg-Lever, A. Rule, R. Schmidt-Ullrich, and W. F. Lever, Ultrastructural localization of in vivo bound immunoglobulin in bullous pemphigoid—a preliminary report. *J. Invest. Dermatol. 64*:47-49 (1975).

123. J. R. Stanley, P. Hawley-Nelson, S. H. Yuspa, E. M. Shevach, and S. I. Katz, Characterization of bullous pemphigoid antigen: A unique basement membrane protein of stratified squamous epithelia. *Cell 24*:897-903 (1981).

124. L. A. Diaz, N. J. Calvanico, T. B. Tomasi, and R. E. Jordon, Bullous pemphigoid antigen: Isolation from normal human skin. *J. Immunol. 118*:455-460 (1977).

125. L. A. Diaz, H. Patel, and N. J. Calvanico, Bullous pemphigoid antigen. II. Isolation from the urine of a patient. *J. Immunol. 122*:605-608 (1979).

126. S. Stenman and A. Vaheri, Distribution of a major connective tissue protein, fibronectin, in normal human tissues. *J. Exp. Med. 147*:1054-1064 (1978).

127. M. Matsuda, N. Yoshida, N. Aoki, and K. Wakabayashi, Distribution of cold-insoluble globulin in plasma and tissues. *Ann. N. Y. Acad. Sci 312*: 56-73 (1978).

128. A. Quaroni, K. J. Isselbacher, and E. Ruoslahti, Fibronectin synthesis by epithelial crypt cells of rat small intestine. *Proc. Natl. Acad. Sci. USA 75*:5548-5553 (1978).

252 CLARK AND COLVIN

129. J. R. Couchman, W. T. Gibson, D. Thom, A. C. Weaver, D. A. Rees, and W. E. Parish, Fibronectin distribution in epithelial and associated tissues of the rat. *Arch. Dermatol. Res. 266*:295-310 (1979).

130. T. Suda, T. Nishida, Y. Ohashi, Y. Nakagawa, and R. Manabe, Fibronectin appears at the site of corneal stromal wound in rabbits. *Curr. Eye Res. 1*:553-556 (1982).

131. R. A. F. Clark, H. J. Winn, H. F. Dvorak, and R. B. Colvin, Fibronectin beneath reepithelializing epidermis in vivo; sources and significance. *J. Invest. Dermatol. (Suppl.) 80*:26S-30S (1983).

132. W. T. Gibson, J. R. Couchman, and A. C. Weaver, Fibronectin distribution during the development of fetal rat skin. *J. Invest. Dermatol. 81*: 480-485 (1983).

133. O. Fyrand, Studies on fibronectin in the skin. I. Indirect immunofluorescence studies in normal human skin, *Br. J. Dermatol. 101*:263-270 (1979).

134. T. D. Oberly, in *Fibronectin* (D. F. Mosher, ed.). Academic Press, New York, in press.

135. L. S. Fujikawa, C. Cintron, C. S. Foster, and R. B. Colvin, Fibronectin in the developing rabbit cornea. *Fed. Proc. 40*:769 (1981).

136. L. S. Fujikawa, M. R. Mandel, D. W. Vastine, C. S. Foster, and R. B. Colvin, Fibronectin, laminin, type IV collagen and bullous pemphigoid antigen in rabbit corneal epithelial wounds in culture. *Invest. Ophthal. Vis. Sci. (Suppl.) 24*:45 (1983).

137. J. M. Foidart, E. W. Bere, Jr., M. Yaar, S. I. Rennard, M. Gullino, G. R. Martin, and S. I. Katz, Distribution and immunoelectron microscopic localization of laminin, a noncollagenous basement membrane glycoprotein. *Lab. Invest. 42*:336-342 (1980).

138. R. Timpl, H. Rohde, P. G. Robey, S. I. Rennard, J. M. Foidart, and G. R. Martin, Laminin—a glycoprotein from basement membranes. *J. Biol. Chem. 254*:9933-9937 (1979).

139. A. E. Chung, R. Jaffe, I. L. Freeman, J. P. Vergnes, J. E. Braginski, and B. Carlin, Properties of a basement membrane-related glycoprotein synthesized in culture by a mouse embryonal carcinoma-derived cell line. *Cell 16*:277-287 (1979).

140. L. Risteli and R. Timpl, Isolation and characterization of pepsin fragments of laminin from human placental and renal basement membranes. *Biochem. J. 193*:749-755 (1981).

141. D. Woodley, D. Sauder, M. J. Talley, M. Silver, G. Grotendorst, and E. Qwarnstrom, Localization of basement membrane components after dermalepidermal junction separation. *J. Invest. Dermatol. 81*:149-153 (1983).

142. J. A. Madri, F. J. Roll, H. Furthmayr, and J. M. Foidart, Ultrastructural localization of fibronectin and laminin in the basement membranes of the murine kidney. *J. Cell Biol. 86*:682-687 (1980).

143. P. A. Courtoy, R. Timpl, and M. G. Farquhar, Comparative distribution of laminin, type IV collagen, and fibronectin in the rat glomerulus. *J. Histochem. Cytochem. 30*:874–886 (1982).

144. J. R. Sanes, Laminin, fibronectin, and collagen in synaptic and extra-synaptic portions of muscle fiber basement membrane. *J. Cell Biol. 93*:442–451 (1982).

145. G. W. Laurie, C. P. Leblond, and G. R. Martin, Localization of type IV collagen, laminin, heparan sulfate proteoglycan, and fibronectin to the basal lamina of basement membranes. *J. Cell Biol. 95*:340–344 (1982).

146. E. D. Hay, Collagen and embryonic development, in *Cell Biology of the Extracellular Matrix* (E. D. Hay, ed.). pp. 379–409. 1981, Plenum Press, New York.

147. V. P. Terranova, D. H. Rohrback, and G. R. Martin, Role of laminin in the attachment of PAM 212 (epithelial) cells to basement membrane collagen. *Cell 22*:719–726 (1980).

148. H. K. Kleinman, D. H. Rohrbach, V. P. Terranova, A. T. Varner, A. T. Hewitt, G. R. Grotendorst, C. M. Wilkes, G. R. Martin, H. Seppa, and E. Schiffman, Collagenous matrices as determinants of cell function. In *Immunochemistry of the Extracellular Matrix, Vol. II*. (H. Furth-mayr, ed.). CRC Press, Boca Raton, Florida, 1982, pp. 151–174.

149. S. P. Sugrue and E. D. Hay, Response of basal epithelial cell surface and cytoskeleton to solubilized extracellular matrix molecules. *J. Cell Biol. 91*:45–54 (1981).

150. B. L. Bender, R. Jaffe, B. Carlin, and A. E. Chung, Immunolocalization of entactin, a sulfated basement membrane component, in rodent tissues, and comparison with GP-2 (laminin). *Am. J. Pathol. 103*:419–426 (1981).

151. J. R. Hassell, P. G. Robey, H. J. Barrach, J. Wilczek, S. I. Rennard, and G. R. Martin, Isolation of a heparan sulfate-containing proteoglycan from basement membrane. *Proc. Natl. Acad. Sci USA 77*:4494–4498 (1980).

152. Y. S. Kanwar, V. C. Hascall, and M. G. Farquhar, Partial characterization of newly synthesized proteoglycans isolated from the glomerular basement membrane. *J. Cell Biol. 90*:527–532 (1981).

153. Y. S. Kanwar and M. G. Farquhar, Isolation of glycosaminoglycans (heparan sulfate) from glomerular basement membranes. *Proc. Natl. Acad. Sci. USA 76*:4493–4497 (1979).

154. M. G. Farquhar, The glomerular basement membrane. A selective macromolecular filter, in *Cell Biology of the Extracellular Matrix* (E. D. Hay, ed.). Plenum Press, New York, 1981, pp. 335–378.

155. J. M. Fitch, E. Gibney, R. D. Sanderson, R. Mayne, and T. F. Linsen-mayer, Domain and basement membrane specificity of a monoclonal antibody against chicken type IV collagen. *J. Cell Biol. 95*:641–647 (1982).

156. H. Yaoita, J. M. Foidart, and S. I. Katz, Localization of the collagenous component in skin basement membrane. *J. Invest. Dermatol. 70*: 191–193 (1978).

157. R. Timpl, H. Wiedemann, V. Van Delden, H. Furthmayr, and K. Kuhn, A network model for the organization of type IV collagen molecules in basement membranes. *Eur. J. Biochem. 120*:203–211 (1981).

158. H. Furthmayr and K. von der Mark, The use of antibodies to connective tissue proteins in studies on their location in tissue, in *Immunochemistry of the Extracellular Matrix, Vol. II*, (H. Furthmayr, ed.), CRC Press, Boca Raton, Florida, 1982, pp. 89–117.

159. F. J. Roll, J. A. Madri, J. Albert, and H. Furthmayr, Codistribution of collagen types IV and AB2 in basement membranes and mesangium of the kidney: An immunoferritin study of ultrathin frozen sections. *J. Cell Biol. 85*:597–616 (1980).

160. K. S. Stenn, J. A. Madri, and F. J. Roll, Migrating epidermis produces AB2 collagen and requires continual collagen synthesis for movement. *Nature 277*:299–232 (1979).

161. S. Gay, T. F. Kresina, R. Gay, E. F. Miller, and L. F. Montes, Immunohistochemical demonstration of basement membrane collagen in normal human skin and in psoriasis. *J. Cutan. Pathol. 6*:91–95 (1979).

162. G. R. Grotendorst, E. Seppa, H. E. Kleinman, and G. R. Martin, Attachment of smooth muscle cells to collagen and their migration toward platelet-derived growth factor. *Proc. Natl. Acad. Sci. USA 78*: 3669–3672 (1981).

163. W. S. Krawczyk, A pattern of epidermal cell migration during wound healing. *J. Cell Biol. 49*:247–263 (1971).

164. H. Hintner, P. O. Fritsch, J. M. Foidart, G. Stingl, G. Schuler, and S. I. Katz, Expression of basement membrane zone antigens at the dermoepibolic junction in organ cultures of human skin. *J. Invest. Dermatol. 74*:200–204 (1980).

165. G. Odland, and R. Ross, Human wound repair. I. Epidermal regeneration. *J Cell. Biol. 39*:135–151 (1968).

166. G. Gabbiani, C. Chaponnier, and I. Huttner, Cytoplasmic filaments and gap junctions in epithelial cells and myofibroblasts during wound healing. *J. Cell Biol. 76*:561–568 (1978).

167. W. S. Krawczyk and G. F. Wilgram, Hemidesmosome and desmosome morphogenesis during epidermal wound healing. *J. Ultrastruct. Res. 45*: 93–101 (1973).

168. N. S. McNutt, Ultrastructural comparison of the interface between epithelium and stroma in the basal cell carcinoma and control human skin. *Lab. Invest. 35*:132–142 (1976).

169. I. I. Singer, Fibronexus-like adhesion sites are present at the invasive zones of human epidermoid carcinomas, in *Fortieth Annual Proceedings of the Electron Microscopy Society of America* (G. W. Bailey, ed.). 1982, pp. 106–109.

170. H. C. Grillo and J. Gross, Collagenolytic activity during mammalian wound repair. *Dev. Biol.* 15:300-317 (1967).
171. R. B. Donoff, J. E. McLennon, and H. C. Grillo, Preparation and properties of collagenases from epithelium and mesenchyme of healing mammalian wounds. *Biochem. Biophys. Acta* 227:639-653 (1971).
172. M. Berman, R. Leary, and J. Gage, Evidence for a role of the plasminogen activator-plasma system in corneal ulceration. *Invest. Ophthalmol. Vis. Sci.* 19:1204-1221 (1980).
173. M. Kubo, D. A. Norris, S. E. Howell, S. T. Ryan, and R. A. F. Clark, Human keratinocytes synthesize, secrete and deposit fibronectin in the pericellular matrix. *J. Invest. Dermatol.* 82:580-586 (1984).
174. K. Alitalo, E. Kuismanen, R. Myllyla, U. Kiistala, S. Askoseljavaara, and A. Vaheri, Extracellular matrix proteins of human epidermal keratinocytes and feeder 3T3 cells. *J. Cell Biol* 94:497-505 (1982).
175. A. L. Kariniemi, V. P. Lehto, T. Vartio, and I. Virtanen, Cytoskeletol and pericellular matrix organization of pure adult human keratinocytes cultured from suction-blister roof epidermis. *J. Cell Sci.* 58:49-61 (1982).
176. K. W. Brown and E. K. Parkinson, Glycoproteins and glycosaminoglycans of cultured normal human epideram keratinocytes. *J. Cell Sci.* 61: 325-338 (1983).
177. K. Alitalo, H. Halila, E. Vesterinen, and A. Vaheri, Endo- and ectocervical human uterine epithelial cells distinguished by fibronectin production and keratinization in culture. *Cancer Res.* 42:1142-1146 (1982).
178. D. L. Nelson, C. D. Little, and G. Balian, Distribution of fibronectin and laminin in basal cell epitheliomas. *J. Invest. Dermatol.* 80:446-452 (1983).
179. R. E. Grimwood, J. C. Huff, J. W. Harbel, and R. A. F. Clark, Fibronectin in basal cell epithelioma: Sources and significance. *J. Invest. Dermatol.* 82:145-149 (1984).
180. E. G. Hayman and E. Ruoslahti, Distribution of fetal bovine serum fibronectin and endogenous rat cell fibronectin in extracellular matrix. *J. Cell Biol.* 83:255-259 (1979).
181. J. R. Stanley, O. M. Alvarez, E. W. Bere, Jr., W. H. Eaglstein, and S. I. Katz, Detection of basement membrane zone antigens during epidermal wound healing in pigs. *J. Invest. Dermatol.* 77:240-243 (1981).
182. D. BenEzra and J. M. Foidart, Collagens and non collagenous proteins in the human eye. I. Corneal stroma in vivo and keratocyte production in vitro. *Curr. Eye Res.* 1:101-110 (1981).
183. M. Kurkinen, K. Alitalo, A. Vaheri, S. Stenman, and L. Saxen, Fibronectin in the development of embryonic chick eye. *Dev. Biol.* 69: 589-600 (1979).
184. U. Ali and R. O. Hynes, Effect of LETS glycoprotein on cell motility. *Cell* 14:439-446 (1978).
185. P. H. Burrill, I. Bernardini, H. K. Kleinman, and N. Kretchmer, Effect

of serum, fibronectin, and laminin on adhesion of rabbit intestinal epithelial cells in culture. *J. Supramol. Struct. Cell Biochem.* *16*:385-392 (1981).

186. B. A. Gilchrest, J. K. Calhoun, and T. Maciag, Attachment and growth of human keratinocytes in a serum-free environment. *J. Cell Physiol.* *112*:197-206 (1982).

187. J. C. Murray, G. Stingl, H. K. Kleinman, G. R. Martin, and S. I. Katz, Epidermal cells adhere preferentially to type IV (basement membrane) collagen. *J. Cell Biol.* *80*:197-202 (1979).

188. W. Federgreen and K. S. Stenn, Fibronectin (LETS) does not support epithelial cell spreading. *J. Invest. Dermatol.* *75*:261-263 (1980).

189. I. Vlodavsky and D. Gospodarowicz, Respective roles of laminin and fibronectin in adhesion of human carcinoma and sarcoma. *Nature 28*: 304-306 cells (1981).

190. R. A. F. Clark, R. L. Wertz, and J. M. Folkvord, Fibronectin, as well as other extracellular matrix proteins, mediate human keratinocyte adherence. *J. Invest. Dermatol.* *85*:378-383 (1985).

191. K. S. Stenn, J. A. Madri, T. Tinghitella, and V. P. Terranova, Multiple mechanisms of dissociated epidermal cell spreading. *J. Cell Biol. 96*: 63-67 (1983).

192. E. Mitrani and R. Marks, Towards characterization of epidermal cell migration promotion activity in serum. *Br. J. Dermatol.* *99*:513-518 (1978).

193. K. S. Stenn, Epibolin: A protein human plasma that supports epithelial cell movement. *Proc. Natl. Acad. Sci. USA 78*:6907-6911 (1981).

194. G. M. Edelman, Cell adhesion molecules. *Science 219*:450-457 (1983).

195. S. Taylor and J. Folkman, Protamine is an inhibitor of angiogenesis. *Nature 297*:307-312 (1982).

196. J. Folkman, E. Merler, C. Abernathy, and G. Williams, Isolation of a tumor factor responsible for angiogenesis. *J. Exp. Med.* *133*:275-288 (1971).

197. J. B. Weiss, R. A. Brown, S. Kumar, and P. Phillips, An angiogenic factor isolated from tumours: A potent low-molecular-weight compound. *Br. J. Cancer 40*:493-496 (1979).

198. A. H. Fenselau, S. Watt, and R. J. Mello, Tumor angiogenic factor. Purification from the Walker 256 rat tumor. *J. Biol. Chem.* *256*:9605-9611 (1981).

199. P. A. D'Amore, B. M. Glaser, S. K. Brunson, and A. H. Fenselau, Angiogenic activity from bovine retina: Partial purification and characterization. *Proc. Natl. Acad. Sci. USA 78*:3068-3072 (1981).

200. R. G. Azizkhan, J. C. Azizkhan, B. R. Zetter, and J. Folkman, Mast cell heparin stimulates migration of capillary endothelial cells in vitro. *J. Exp. Med.* *152*:931-944 (1980).

201. M. J. Banda, D. R. Knighton, T. K. Hunt, and Z. Werb, Isolation of a

nonmitogenic angiogenesis factor from wound fluid. *Proc. Natl. Acad. Sci. 79*:7773-7777 (1982).

202. M. M. Sholley, M. A. Gimbrone, and R. S. Cotran, Cellular migration and replication in endothelial regeneration. A study using irradiated endothelial cultures. *Lab. Invest. 36*:18-25 (1977).

203. D. H. Ausprunk and J. Folkman, Migration and proliferation of endothelial cells in preformed and newly formed blood vessels during tumor angiogenesis. *Microvasc. Res. 14:*53-65 (1977).

204. J. A. Madri, Endothelial cell-matrix interactions in hemostatis. *Prog. Hemost. Thromb. 6*:1-24 (1982).

205. E. A. Jaffe and D. F. Mosher, Synthesis of fibronectin by cultured endothelial cells. *J. Exp. Med. 147*:1779-1791 (1978).

206. E. J. Macarak, E. Kirby, T. Kirk, and N. A. Kefalides, Synthesis of cold-insoluble globulin by cultured calf endothelial cells. *Proc. Natl. Acad. Sci. 75*:2621-2625 (1978).

207. C. R. Birdwell, D. Gospodarowicz, and G. L. Nicolson, Identification, localization, and role of fibronectin in cultured bovine endothelial cells. *Proc. Natl. Acad. Sci. USA 75*:3273-3277 (1978).

208. A. Palotie, K. Tryggvason, L. Peltonen, and H. Seppa, Components of subendothelial aorta basement membrane. Immunohistochemical localization and role in cell attachment. *Lab. Invest. 49*:362-370 (1983).

209. R. A. F. Clark, J. Folkvord, P. Kern, and R. Eckel, Human endothelial cells from both large and small blood vessels demonstrate fibronectin dependent adherence. *J. Cell Biol. 97*:327a (1983).

210. P. Davison and M. Karasek, Serial cultivation of human dermal vessel endothelium: Role of serum and fibronectin. *Clin. Res. 28*:566A (1980).

211. B. M. Golub, C. S. Foster, and R. B. Colvin, Fibronectin, laminin, type IV collagen, factor VIII and fibrin: Sequential analysis during experimental corneal neovascularization in the guinea pig. *Invest. Ophthal. Vis. Sci. (Suppl.) 22*:27 (1982).

212. B. R. McAuslan, G. N. Hannan, W. Reilly, and F. H. C. Stewart, Variant endothelial cells. Fibronectin as a transducer of signals for migration and neovascularisation. *J. Cell Physiol. 104*:177-186 (1980).

213. B. R. McAuslan, G. N. Hannan, and W. Reilly, Signals causing change in morphologic phenotype, growth mode, and gene expression of vascular endothelial cells. *J. Cell Physiol. 112*:96-106 (1982).

214. P. J. McKeown-Longo and D. F. Mosher, Binding of plasma fibronectin to cell layers of human skin fibroblasts. *J. Cell Biol. 97*:466-472 (1983).

215. R. B. Colvin, P. I. Garner, R. O. Roblin, E. L. Verderber, J. M. Lanigan, and M. W. Mosesson, Cell surface fibrinogen-fibrin receptors on cultured human fibroblasts. Association with fibronectin (cold insoluble globulin LETS protein) and loss in SV40 transformed cells. *Lab. Invest. 41*:464-473 (1979).

216. E. Oh, M. Pierschbacher, and E. Ruoslahti, Deposition of plasma fibro-nectin in tissues. *Proc. Natl. Acad. Sci USA 78*:3218–3221 (1981).

217. L. B. Chen, R. C. Gudor, T. T. Sun, A. B. Chen, and M. W. Mosesson, Control of a cell surface major glycoprotein by epidermal growth fac-or. *Science 197*:776–778 (1977).

218. A. K. C. Li, M. J. Koroly, M. E. Shattenkerk, R. A. Malt, and M. Young, Nerve growth factor: Acceleration of the rate of wound healing in mice. *Proc. Natl. Acad. Sci. USA 77*:4379–4381 (1980).

219. J. M. Hutson, M. Niall, D. Evans, and R. Fowler, Effect of salivary glands on wound contraction in mice. *Nature 279*:793–795 (1979).

220. J. P. Pennypacker, J. R. Hassell, K. M. Yamada, and R. M. Pratt, The influence of an adhesive cell surface protein on chondrogenic expres-sion in vitro. *Exp. Cell Res. 121*:411–415 (1979).

221. C. M. West, R. Lanza, J. Rosenbloom, M. Lowe, H. Holtzer, and N. Avdalovic, Fibronectin alters the phenotypic properties of cultured chick embryo chondroblasts. *Cell 17*:491–501 (1979).

222. R. O. Hynes and A. Destree, Extensive disulfide bonding at the mam-malian cell surface. *Proc. Natl. Acad. Sci. USA 74*:2855–2859 (1977).

223. D. F. Mosher and R. B. Johnson, In vitro formation of disulfide-bonded fibronectin multimers. *J. Biol. Chem., 258*:6595–6601 (1983).

224. E. A. Jaffe, J. T. Ruggiero, L. L. K. Leung, M. J. Doyle, P. J. McKeown-Longo, and D. F. Mosher, Cultured human fibroblasts synthesize and secrete thrombospondin and incorporate it into extracellular matrix. *Proc. Natl. Acad. Sci. USA 80*:998–1002 (1983).

225. S. B. Carter, Principles of cell motility: The direction of cell movement and cancer invasion. *Nature 208*:1183–1187 (1965).

226. V. Gauss-Muller, H. K. Kleinman, G. R. Martin, and E. Schiffman, Role of attachment factors and attractants in fibroblast chemotaxis. *J. Lab. Clin. Med. 96*:1071–1080 (1980).

227. Y. Tsukamoto, W. E. Helsel, and S. M. Wahl, Macrophage production of fibronectin, a chemoattractant for fibroblasts. *J. Immunol. 127*:673–678 (1981).

228. P. Hsieh and L. B. Chen, Behavior of cells seeded in isolated fibronectin matrices. *J. Cell Biol. 96*:1208–1217 (1983).

229. J. A. McDonald, R. M. Senior, G. L. Griffin, T. J. Broekelmann, and P. Prevedel, Relationship between fibronectin mediated fibroblast chemotaxis and cell adhesion. *J. Cell Biol. 95*:123a (1982).

230. R. B. Colvin and R. L. Kradin, The biological activity of fibronectin controlled by alterations in its molecular form? Hypothesis and review. *Surv. Synth. Pathol. Res. 2*:10–20 (1983).

231. G. Majno, The story of the myofibroblasts. *Am. J. Surg. Pathol. 3*: 535–542 (1979).

232. G. Gabbiani, The role of contractile proteins in wound healing and fi-brocontractive disease. *Meth. Achiev. Exp. Pathol. 9*:187–206 (1979).

233. G. B. Ryan, W. J. Cliff, G. Gabbiani, C. Irle, D. Montandon, P. R. Stat-kov, and G. Majno, Myofibroblasts in human granulation tissue. *Hum. Pathol.* 5:55-67 (1974).

234. G. Gabbiani, B. J. Hirschel, G. B. Ryan, P. R. Statkov, and G. Majno, Granulation tissue as a contractile organ. A study of structure and function. *J. Exp. Med.* 135:719-734 (1972).

235. G. Majno, G. Gabbiani, B. J. Hirschel, G. B. Ryan, and P. R. Statkov, Contraction of granulation tissue in vitro: Similarity to smooth muscle. *Science* 173:548-550 (1971).

236. I. L. Singer, The fibronexus: A transmembrane association of fibronectin-containing fibers and bundles of 5 nm microfilaments in hamster and human fibroblasts. *Cell* 16:675-685 (1979).

237. L. T. Furcht, G. Wendelschafer-Crabb, D. F. Mosher, and J. M. Foidart, An axial periodic fibrillar arrangement of antigenic determinants for fibronectin and procollagen on ascorbate treated human fibroblasts. *J. Supramol. Struct.* 13:15-33 (1980).

238. I. I. Singer and P. R. Paradiso, A transmembrane relationship between fibronectin and vinculin (130kd protein): Serum modulation in normal and transformed hamster fibroblasts. *Cell* 24:481-492 (1981).

239. R. O. Hynes and A. T. Destree, Relationships between fibronectin (LETS protein) and actin. *Cell* 15:875-886 (1978).

240. M. H. Heggeness, J. F. Ash, and S. J. Singer, Transmembrane linkage of fibronectin to intracellular actin-containing filaments in cultured human fibroblasts. *Ann. N. Y. Acad. Sci.* 312:414-417 (1978).

241. I. I. Singer, D. W. Kawka, D. M. Kazazis, and R. A. F. Clark, The in vivo codistribution of fibronectin and actin fibers in granulation tissue. Immunofluorescence and electron microscopic studies of the fibronexus at the myofibroblasts surface. *J. Cell. Biol.* 98:2091-2106 (1984).

242. E. Bell, B. Ivarsson, and C. Merrill, Production of a tissue-like structure by contraction of collagen lattices by human fibroblasts of different proliferative potential in vitro. *Proc. Natl. Acad. Sci. USA* 76:1274-1278 (1979).

243. E. Bell, H. P. Ehrlich, D. J. Buttle, and T. Nakatsuji, Living tissue formed in vitro and accepted as skin-equivalent tissue of full thickness. *Science* 211:1052-1054 (1981).

244. E. Bell, S. Sher, B. Hull, C. Merrill, S. Rosen, A. Champson, D. Asselineau, L. Dubertret, B. Coulomb, C. Lapiere, B. Nusgens, and Y. Neveux, The reconstitution of living skin. *J. Invest. Dermatol. (Suppl.)* 81:2-10 (1983).

245. E. G. Hayman, A. Oldberg, G. E. Martin, and E. Ruoslahti, Codistribution of heparan sulfate proteoglycan, laminin, and fibronectin in the extracellular matrix of normal rat kidney cells and their coordinate absence in transformed cells. *J. Cell Biol.* 94:28-35 (1982).

246. K. Hedman, S. Johansson, T. Vartio, L. Kjellen, A. Vaheri, and M.

Hook, Structure of the pericellular matrix: Association of heparan and chondroitin sulfates with fibronectin-procollagen fibers. *Cell 28*:663–671 (1982).

247. P. Bornstein and T. F. Ash, Cell surface-associated structural proteins in connective tissue cells. *Proc. Natl. Acad. Sci. USA 74*:2480-2484 (1977).
248. A. Vaheri, M. Kurkinen, V. P. Lehto, E. Linder, and R. Timpl, Codistribution of pericellular matrix proteins in cultured fibroblasts and loss in transformation: Fibronectin and procollagen. *Proc. Natl. Acad. Sci. 75*:4944-4948 (1978).
249. J. A. McDonald, D. G. Kelley, and T. J. Broekelmann, Role of fibronection in collagen deposition: Fab' to the gelatin-binding domain of fibronectin inhibits both fibronectin and collagen organization in fibroblast extracellular matrix. *J. Cell Biol. 92*:485-492 (1982).
250. L. B. Chen, A. Murray, R. A. Segal, A. Bushnell, and M. L. Walsh, Studies on intercellular LETS glycoprotein matrices. *Cell 14*:377-391 (1978).
251. R. Fleischmajer, W. Dessau, R. Timpl, T. Kreig, C. Luderschmidt, and M. Wiestner, Immunofluorescence analysis of collagen, fibronectin, and basement membrant protein in scleroderma skin. *J. Invest. Dermatol. 75*:270-274 (1980).
252. E. H. Epstein, [α1(III)]₃ Human skin collagen. *J. Biol. Chem. 249*: 3225-3231 (1974).
253. J. N. Clore, I. K. Cohen, and R. F. Diegelmann, Quantitation of collagen types I and III during wound healing in rat skin. *Proc. Soc. Exp. Biol. Med. 161*:337-340 (1979).
254. S. Gay, J. Viljanto, J. Raekallio, and R. Penttinen, Collagen types in early phases of wound healing in children. *Acta Chir. Scand. 144*:205-211 (1978).
255. D. J. Unsworth, D. L. Scott, T. J. Almond, H. K. Beard, E. J. Holborrow, and K. W. Walton, Studies on reticulin. I. Serological and immunohistological investigations of the occurrence of collagen type III, fibronectin and the non-collagenous glycoprotein of Pras and Glynn in reticulin. *Br. J. Exp. Pathol. 63*:154-166 (1982).
256. C. W. Kischer and M. J. C. Hendrix, Fibronectin (FN) in hypertrophic scars and keloids. *Cell Tissue Res. 231*:29-37 (1983).
257. S. M. Cooper, A. J. Keyser, A. D. Beaulieu, E. Ruoslahti, M. E. Nimni, and F. P. Quismorio, Jr., Increase in fibronectin in the deep dermis of involved skin in progressive systemic sclerosis. *Arthritis Rheum. 22*: 983-987 (1979).
258. E. Hahn, G. Wick, D. Pencev, and R. Timpl, Distribution of basement membrane proteins in normal and fibrotic human liver: Collagen type IV, laminin, and fibronectin. *Gut 21*:63-71 (1980).
259. E. Engvall, E. Ruoslahti, and E. J. Miller, Affinity of fibronectin to collagens of different genetic types and to fibrinogen. *J. Exp. Med. 147*:1584-1595 (1978).

260. F. Jilek and H. Hormann, Cold-insoluble globulin (fibronectin). IV. Affinity to soluble collagen of various types. *Hoppe-Seylers Z. Physiol. Chem.* *359*:247-250 (1978).

261. S. Johansson and M. Höök, Heparin enhances the rate of binding of fibronectin to collagen. *Biochem. J.* *187*:521-524 (1980).

262. E. Ruoslahti and E. Engvall, Complexing of fibronectin, glycosaminoglycans and collagen. *Biochim. Biophys. Acta* *631*:350-358 (1980).

263. R. A. Proctor, E. Pendergast, and D. F. Mosher, Fibronectin mediates attachment of *Staphylococcus aureus* to human neutrophils. *Blood* *59*:681-687 (1982).

264. R. T. Wall, S. L. Cooper, and J. C. Kosek, The influence of exogenous fibronectin on blood granulocyte adherence to vascular endothelium in vitro. *Exp. Cell Res.* *140*:205-109 (1982).

265. A. F. Brown and J. M. Lackie, Fibronectin and collagen inhibit cell-substratum adhesion of neutrophil granulocytes. *Exp. Cell Res.* *136*: 225-231 (1981).

266. S. T. Hoffstein, G. Weissman, and E. Pearlstein, Fibronectin is a component of the surface coat of human neutrophils. *J. Cell Sci.* *50*:315-327 (1981).

267. T. Vartio, H. Seppa, and A. Vaheri, Susceptibility of soluble and matrix fibronectins to degradation by tissue proteinases, mast cell chymase and cathepsin. *J. Biol. Chem.* *256*:471-477 (1981).

268. J. Sasaki, M. Imanaka, S. Watanabe, N. Otsuka, and K. Sugiyama, Immuno-electron microscopic localization of fibronectin on rat mast cells. *Experientia.* *38*:495-496 (1982).

269. K. Sugiyama, O. Kamata, and N. Katsuda, Fibronectin on the surface of rat mast cells. *Exp. Cell Res.* *133*:449-452 (1981).

270. R. M. Akers, D. F. Mosher, and J. E. Lilien, Promotion of retinal neuite outgrowth by substratum-bound fibronectin. *Dev. Biol.* *8*:179-188 (1981).

271. S. T. Carbonetto, M. M. Gruver, and D. C. Turner, Nerve fiber growth on defined hydrogel substrates. *Science* *216*:897-899 (1982).

272. J. B. McCarthy, S. L. Palm, and L. T. Furcht, Migration by haptotaxis of a Schwann cell tumor line to the basement membrane glycoprotein laminin. *J. Cell Biol.* *97*:772-777 (1983).

273. A. Baron-Van Evercooren, H. K. Kleinman, H. E. J. Seppa, B. Rentier, and M. Dubois-Dalco, Fibronectin promotes rat Schwann cell growth and motility. *J. Cell Biol.* *93*:211-216 (1982).

274. R. A. Rovasio, A. Delouvee, K. M. Yamada, R. Timpl, and J. P. Thiery, Neutral crest cell migration: Requirements for exogenous fibronectin and high cell density. *J. Cell Biol.* *96*:462-473 (1983).

275. J. Z. Young and P. B. Medawar, Fibrin suture of peripheral nerves. Measurment of the rate of regeneration. *Lancet* *2*:126-128 (1940).

276. H. Matras, The use of fibrin sealant in oral and maxillofacial surgery.

J. Oral Maxillofac. Surg. 40:617–622 (1982).

277. T. R. Podleski, I. Greenberg, J. Schlessinger, and K. M. Yamada, Fibronectin delays the fusion of L6 myoblasts. *Exp. Cell Res. 122*:317–326 (1979).

278. M. A. Arneson, D. E. Hammerschmidt, L. T. Furcht, and R. A. King, A new form of Ehlers-Danlos syndrome. Fibronectin corrects defective platelet function. *JAMA 24*:144–147 (1980).

279. J. L. Weiss, L. T. Furcht, J. D. Cameron, J. D. Nelson, R. G. Lindstron, and D. J. Doughman, Enahnced corneal epithelial wound healing using topical fibronectin, *Invest. Opthalmol. Vis. Sci. (Suppl.) 22*:23 (1982).

280. T. Nishida, S. Nakagawa, T. Awata, Y. Ohashi, K. Watanabe, and R. Manabe, Fibronectin promotes epithelial migrates of cultured rabbit cornea in situ. *J. Cell Biol. 97*:1653–1657 (1983).

281. P. O. Ganrot, Variation of the concentrations of some plasma proteins in normal adults, in pregnant women and in newborns. *Scand. J. Clin. Lab. Invest. (Suppl.) 124*:83–88 (1972).

282. J. K. Czop, J. L. Kadish, and K. F. Austen, Purification and characterization of a protein with fibronectin determinants and phagocytosis-enhancing activity. *J. Immunol. 129*:163–167 (1982).

283. F. Grinnell and D. Minter, Attachment and spreading of baby hamster kidney cells to collagen substrata: Effects of cold-insoluble globulin *Proc. Natl. Acad. Sci. USA 75*:4408–4412 (1978).

284. H. Metzger, The affect of antigen on antibodies: Recent studies. *Contemp. Topic Molec. Immunol. 9*:119–152 (1978).

285. C. G. Cochrane and J. H. Griffin, The biochemistry and pathophysiology of the contact system of plasma. *Adv. Immunol. 33*:241–306 (1982).

286. A. P. Kaplan, Hageman factor-dependent pathways: mechanism of initiation and bradykinin formation. *Fed. Proc. 42*:3123–3127 (1983).

Index

Milton Keynes UK
Ingram Content Group UK Ltd.
UKHW020024071024
449327UK00032B/2918